Pablo Carbonell

Metabolic Pathway Design

A Practical Guide

 Springer

Pablo Carbonell
Manchester Institute of Biotechnology
University of Manchester
Manchester, UK

ISSN 2509-6125   ISSN 2509-6133   (electronic)
ISBN 978-3-030-29864-7   ISBN 978-3-030-29865-4   (eBook)
https://doi.org/10.1007/978-3-030-29865-4

This Springer imprint is published by the registered company Springer Nature Switzerland AG
The registered company address is: Gewerbestrasse 11, 6330 Cham, Switzerland

Many are stubborn in pursuit of the path they have chosen, few in pursuit of the goal.
Friedrich Nietzsche

Preface

It was a cold midday at the turn of the millennium. I emerged from the gloomy desk of my basement lab to enjoy a sunny moment at the Brooklyn Polytech plaza. Next to me a massive building from the Bell Telephone Labs rose as an unavoidable sight. There it was where many great engineering things began, where systems, circuits, and devices were designed, built, and tested.

Closing the loop between engineering and biology has been since then part of my research. This is the main goal of this textbook. Here, I would like to provide you with the tools that we use nowadays in order to engineer and design genetic circuits.

There at Brooklyn Polytech, I worked with Prof. Zhong-Ping Jiang, a mathematician expert in control engineering. As we walked across the plaza, he challenged me as what was going to be the technology that would shape the new century. Fields such as electronics and computer science were the defining technology of our past century. Nanotechnologies were certainly opening new frontiers. But Prof. Jiang was clear about it: "Pablo, this next century is the century of engineering biology and biotechnology."

Those words still resound on me. For the first time I realized that our research in automation and machine learning at the Brooklyn basement lab was part of a fundamental change taking place in the field of biotechnology. Modern biotechnology is rooted in automation, modeling, and design.

The goal of this textbook is to invite you into the exciting field of engineering biology. If you feel as I do that bioengineering design is our next revolution, please come along, be open, and enjoy this book as your special learning journey into biological engineering design.

Designed in Manchester, San Francisco, and Alcoy
Pablo Carbonell December 2018.

Acknowledgments

This textbook is not a thought experiment. This book stems from fruitful discussions and arduous dry and wet lab experiments throughout the years.

This book is infused with the amazing work of Manchester's Synbiochem core team. A great thanks to Adrian J Jervis, Andrew Currin, Barbara Ribeiro, Christopher J Robinson, Cunyu Yan, Katherine A Hollywood, Mark Dunstan, Maria Vinaixa, Neil Swainston, Rob Meckin, Reynard Spiess, Rehana Sung, and Sandra Taylor. Thanks to all the great people at the Manchester Institute of Biotechnology and The University of Manchester for their inspiring visions about the incoming industrial biomanufacturing revolution. Many thanks to Nigel S. Scrutton, Eriko Takano, Nicholas J Turner, Perdita Barran, Rainer Breitling, Rosalind le Feuvre, Carole Goble, Royston Goodacre, Douglas B. Kell, Phil Shapira, and Jason Micklefield.

Deep thanks to Jean-Loup Faulon and his great team at the INRA, this book is rooted in our shared engineering approach to metabolic engineering. Thank you Anne-Gaëlle Planson, Baudoin Delépine, Claire Baudier, Cyrille Pauthenier, Davide Fichera, Ioana Grigoras, Pierre Parutto, Shashi Pandit, Tamás Fehér, and Thomas Duigou.

A warm thanks to my friends and colleagues Francis Planes at the Universidad de Navarra, Jesus Picó at the Universitat Politècnica de València, Hector Garcia-Martín at the Joint BioEnergy Institute, Angel Goñi at Newcastle University, Maria Miteva at the Université Paris Diderot, Ferran Sanz at the Universitat Pompeu Fabra, Jean-Yves Trosset at Sup'Biotech for their inspiring discussions and views.

Finally, I would like to acknowledge Amrei Strehl at Springer Nature for her support and leading role in the Learning Materials in Biosciences series and Sybille Czerniakowski for her valuable suggestions and feedback.

I am probably missing a lot of other persons who were part of the inception of this textbook. Synthetic biology is a community-based effort. Without their input, many of the achievements described in this book would have never been possible. Thanks to you all.

Contents

III   Metabolic Pathway Design

Metabolic Pathway Modeling

Contents

Getting on the Path to Engineering Biology

© Springer Nature Switzerland AG 2019
P. Carbonell, *Metabolic Pathway Design*, Learning Materials in Biosciences,
https://doi.org/10.1007/978-3-030-29865-4_1

1

What You Will Learn in This Chapter

A new form of engineering biology has become possible. Rising as one of the most exciting scientific achievements of the last decade, this technology has come a long way from its early starts as garage biology into defining a novel paradigm in biotechnology. Nowadays, synthetic biology is a burgeoning technology bringing together biologists, biochemists, computer scientists, physicists, industrial engineers, economists and sociologists. Metabolic pathway design has been at the core of such revolution. Pathway designers envisioned a new discipline where labs exchanged biological circuits through the cloud; where labs embraced automation and artificial intelligence as their core technologies. You are welcome to join this new biotechnology revolution. Your role as a metabolic pathway designer is to master bio-CAD tools in order to rigorously explore the design space of bio-based chemical production. To do that, you need first to understand the basic principles of modeling, simulation and optimization, the basic functioning of the cell, the main concepts of systems and synthetic biology, as well as the basic protocols of biotechnology. As a first introduction, this chapter provides an overview of the design principles of automated synthetic biology as well as the perspectives that such new technology will bring in for all of us.

1.1 Synthetic Biology Platforms for Industrial Biomanufacturing

The future of biotechnology is extremely bright. Biology on its own has undergone an amazing revolution. Today, the bio-based economy is a key part of the 4th Industrial Revolution along with automation, robotics, artificial intelligence and big data [8]. Recent technological advances are allowing scientists to redesign organisms or even build new life forms from scratch. This new developing area of research steps beyond traditional genetic modification to combine science and engineering in what is known as **synthetic biology** by means of introducing new technologies for engineering biology.

The synthetic biology world market will reach \$40 billion by 2020 globally [9]. In this rising new bioeconomy, bio-based production of chemicals and materials are alternatives to chemical processes both economically viable and ecologically sustainable [5]. Nature rather than artificial processes singles out as the main chemical source of diversity that drives next generation industrial biotechnology. However, harnessing the power of natural products for the new bioeconomy will require of a good understanding of biosynthesis processes and the ability of developing predictive models of bioactivity for downstream products. In next generation biomanufacturing, modeling, artificial intelligence and big data analysis will become essential in order to optimize the processes. Ultimately, synthetic biology should be able to model a biological system so that it is possible to predict exactly what needs to be re-programmed in order to achieve the desired behaviour. In working towards that goal, synthetic biology will contribute to our understanding of biology as well as providing products such as novel materials, biosensors, medicines and therapies [11].

Engineering biology thus involves not only understanding the basics of genetic engineering and systems and synthetic biology, but also mastering the concepts and tools from **modeling and design** that are described in this textbook. This first part starts by introducing the **Design-Build-Test-Learn cycle** associated with modern engineering approaches to biology. Essential tools for pathway design such as those for modeling biological systems and chemistry-based models will be provided. Using these basic tools, the second part will focus on describing computational protocols for metabolic pathway design, from enzyme selection to pathway discovery and enumeration. The final part of the book will put metabolic pathway design in the context of industrial biotechnology to understand the challenges associated with pathway optimization. Current technological solutions will be discussed with a focus on experimental design and machine learning solutions.

1.2 Automating the Design-Build-Test-Learn Cycle

In spite of the promising future perspectives, industrial biotechnology processes still remain costly and few bio-based products have reached the market. In order to transition synthetic biology and metabolic engineering into a biomanufacturing technology the development of an **automated pipeline** is necessary [3]. An automated pipeline for engineering biology comprises bioproduction pathway design tools (**Design**), robotized strain engineering (**Build**), high-throughput product quantification (**Test**), and data analysis and redesign through machine learning (**Learn**) [2] (see ◻ Fig. 1.1).

In the **Design stage**, starting from a target compound of interest, algorithms are used to explore known and putative novel biochemical routes connecting the target chemical to the host organism. Sequences encoding enzymes catalyzing each reaction in the pathway are then identified and combined with regulatory elements. Smart optimal designs of combinatorial libraries are selected in order to explore relationships between design factors.

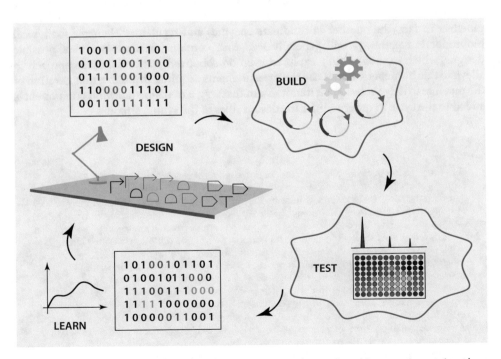

◻ **Fig. 1.1** The role of the metabolic pathway designer is to select and combine genetic parts in order to obtain the desired behavior once assembled in the chassis strain. The design blueprints are sent to the build and test platforms in the cloud. Experimental results are analyzed in order to refine the models and define the design rules for the next pipeline iteration

In the **Build stage**, the pipeline proceeds through a robotized platform for gene cloning where the selected DNA parts are inserted into a plasmid vector and introduced into a host organism such as *Escherichia coli* for replication, selection and screening (see ▶ Box 1.1). Parts are then combined through one of the several pathway assembly technologies that are available, such as Golden Gate, Gibson, Ligase Cycling Reaction, etc [4]. Strain transformation is the process by which the recombinant DNA is inserted into the chassis host. Resulting bacterial cultures are then grown in selected media conditions. If the pathway contains inducible elements, inducers are added into the media at selected induction times. Cell-free systems provide an alternative approach where single functional DNA parts are expressed in chassis strains which are broken down in a controlled way (lysed) and mixed to assemble the pathway [6].

In the **Test stage**, samples are prepared for processing through automated high-throughput screening. Several technologies exist to perform large-scale quantification of the cell response, such as proteomics (protein concentrations), transcriptomics (protein expression levels), metabolomics (metabolite concentrations), etc. Metabolomics is one essential part of the pipeline in order to quantify the titers for the target chemicals and key intermediates.

In the **Learn stage**, quantified experimental data from the grown combinatorial libraries at different conditions are analyzed in order to identify significant effects and hidden relationships between the designs and the resulting responses. Predictive models are developed in order to infer design rules for the next iteration of the cycle.

Several challenges are still present. Notably, the complexity associated with the large combinatorial design space appears as one of the challenges in order to streamline the pipeline. In fact, the number of candidate enzymes and regulatory elements for typical bioproduction pathways will generally lead to a combinatorial explosion of possible designs, which cannot be fully explored even on robotized platforms. One approach to alleviate this bottleneck would use the measurements acquired through a first iteration of the pipeline to drive subsequent iterations. In that way, a **machine learning** component is added to the cycle in order to infer the design rules of the synbio circuits.

Box 1.1

Cells carry the information for protein production stored in their DNA genetic code. A cell expresses the DNA in order to produce a protein. Transcription is the process that consists in decoding the DNA sequence into an intermediary messenger RNA (mRNA). Transcription starts when a RNA polymerase binds to some specific DNA region. Transcription can occur constitutively without any intervention or be regulated by some transcription factor. Inducers are molecules that can serve to activate or represse the transcription factor. The mRNA is translated into a protein in a process known as translation, which occurs when ribosomes attach the mRNA. Ribosomes recognize specific sequence regions called ribosome binding sites (RBS), which will determine the rate of protein production.

Plasmids are circular DNA structures that can be used to introduce and express foreign DNA into organisms such as bacteria. Bacterial strains naturally carry plasmids conferring antibiotic resistance with the ability to replicate as autonomous genetic elements. They appear in the cell in as many copies as their plasmid copy number. These vector backbones can be cut and edit in order to insert desired DNA parts through ligation. Transformation is the process of inserting new plasmids carrying DNA parts into competent cells, i.e. cells that are chemically or electrically modified to make them able to take up exogenous DNA. The new cell is called a recombinant cell.

1.3 Scaling Up and Down Industrial Bioprocesses

Automated Design-Build-Test-Learn pipelines allow **rapid prototyping** of pathway designs by growing and testing samples of different cultures. However, selected prototypes need to be portable. Proof-of-concepts need to be scaled-up to achieve optimal scale at industrial levels. The life cycle of biotechnology products involves working at different scales. **Scaling-up** often consists in promoting a promising prototype found at the wet-lab from the well-plate into shaking flasks and ultimately into bioreactors. However, what was observed at the bottom of the scale would not necessarily behave as such at the top. Due to such principle, the scaling-up of a process will often need to go through a tortuous process in order to move from a proof-of-concept into the final product. A transition period prone to failures.

Engineering biology faces both scale up and scale down challenges [7]. The conditions of the scaled-down lab assays would not necessarily reproduce those of the scaled-up fermenter process. In order to lower the risks of such approach, optimization of the scaling-up step should be considered right from the beginning of the design. Alternatively, **scaling-down** consists in mimicking large scale conditions using laboratory-scale systems. The basic principle of scaling down is to bring the strains under the same conditions as those found in a bioreactor or a fermenter. For instance by switching between glucose feeding and glucose-limited conditions, the response of the system under different conditions can be identified, modeled, and characterized.

1.4 Agile Automated Biodesign, Cloud Labs and Artificial Intelligence

As in many other manufacturing applications such as in the aerospace, electronics or automotive industries, **computer-aided design** (CAD) for biomanufacturing allows increasing the productivity of the designer and the quality of designs. CAD for biomanufacturing replaces traditional trial-and-error approaches in biological engineering with computer systems to assist in the creation, modification, analysis, and optimization of biological designs. Automated biodesign combines CAD systems with engineering principles and automated biomanufacturing [1] to augment human capabilities, streamlining in that way the transfer of a biological design into wet-lab experiments [10]. Modern biomanufacturing has adopted an **agile approach** in order to improve the efficiency and flexibility of the Design-Build-Test-Learn cycle. Rather than keeping a fixed set of design recipes, agile biodesign requires of continuous upgrades based on build, test and learn interactions.

Physical wet-labs with standardized protocols and automated robotic platforms are increasingly becoming remote bench desk platforms or **cloud labs** for the prototyping of biological designs. Cloud labs allow researchers to conduct experiments via remote robotic controls, virtualizing the wet-lab. Cloud biomanufacturing is transforming biological engineering into biomanufacturing services allowing intelligent sharing of resources. **Automation** is making possible this new paradigm.

Artificial intelligence is a core technology enabling such transition from biological engineering into industrial biomanufacturing. Next big ideas and opportunities in engi-

1

neering biology for industrial biotechnology will arise from automated learning and design. Machine learning will lead the next round of innovation through automated generation of design blueprints, cloud manufacturing instructions, and test analysis of target marketable biological circuits, e.g., those for bioproduction of fine chemicals, biologics, biosensors or advanced biomaterials.

1.5   Metabolic Pathway Design in the Biofoundry

The daily work of a metabolic pathway designer in a biofoundry does not differ much from those of designers working on architectural, mechanical or electronics design. Several biological engineering interrelated tasks are part of the work of the designer (see ◘ Fig. 1.2). As a metabolic pathway designer, you will be often involved in identifying promising chemical targets, perhaps because the molecule is a building block for some novel polymer or because it is a novel antibiotic with great potency and specificity. Similarly, your work will consist on the discovery of the biochemical steps that would make possible producing the target chemical in a host chassis organism, such as a bacteria, a yeast, etc. You are also expected to be able to select the enzyme gene sequences that are candidates to catalyze the steps in the pathways by mining biochemical information and diversity from organisms such as plants, extremophiles or fungi. Your next

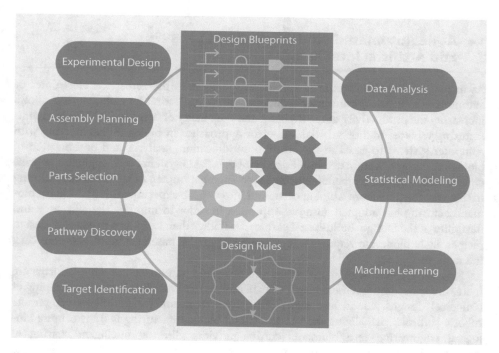

◘ **Fig. 1.2**   Metabolic pathway design starts with the identification of chemical targets followed by pathway discovery and genetic parts selection. Assembly planning of the designated parts is performed in order to generate an experimental design. Resulting design blueprints are transferred to the product pipeline. The experimental data are analyzed in order to build statistical and machine learning-based models. Their design rules will guide the next pathway design iteration

task will be to collect the different biological parts, gene sequences, regulatory elements and to decide how they will be combined in the host chassis in order to efficiently optimize the design.

Once the design project achieves the maturity stage, i.e., the desired combination of genetic parts has been selected, **design blueprints** containing the combinatorial information of the genetic library are generated. Such blueprints are encoded in some standard representation and then transferred from the Design to the Build stage. The Build stage will implement your design by converting the combinatorial library information into a set of instructions or work list to be carried out by the robotics platform in the lab. Interestingly, the actual assembly of your genetic construct, plasmids and strains can be carried out in a cloud lab, i.e., in a highly automated synthetic biology lab operated remotely through robotic instructions. Once your combinatorial library of strains is built, they are transferred to the Test stage, where cultures are grown, samples are analyzed and products quantified. Again, such Test operations can be performed in an automated facility that is physically located in a separate space from the Build facilities following in that way the **cloud biomanufacturing** paradigm.

The experimental data quantified at the Test stage are transferred to the Learn platform. The metabolic pathway designer can play a key role at this stage in order to analyze the data and to build a model that relates the different design factors with the observed behavior. Statistical and machine learning-based analysis are carried out in the cloud computational platform and predictions are used to refine the model. Main factors having a significant impact in the response are identified among the selected parts such as promoters, plasmids, genes, etc. This valuable information is used in order to infer **design rules**. Such rules are transferred into the Design stage in order to select a redesigned optimal combinatorial library as a starting step for the next iteration of the Design-Build-Test-Learn cycle.

Take Home Message

- The bio-based economy is a key part of the 4th Industrial Revolution along with automation, robotics, artificial intelligence and big data.
- The automated pipeline for engineering biology is based on the Design-Build-Test-Learn cycle. Cloud labs and automation are part of the new paradigm.
- The life cycle of biotechnology products involves working at different scales from wet-lab to industrial levels.
- Your role as a metabolic pathway designer is to master bio-CAD tools in order to rigorously explore the design space of bio-based chemical production.

References

1. Appleton, E., Densmore, D., Madsen, C., Roehner, N.: Needs and opportunities in bio-design automation: four areas for focus. Curr. Opin. Chem. Biol. **40**, 111–118 (2017). https://doi.org/10.1016/j.cbpa.2017.08.005
2. Carbonell, P., Currin, A., Jervis, A.J., Rattray, N.J.W., Swainston, N., Yan, C., Takano, E., Breitling, R.: Bioinformatics for the synthetic biology of natural products: integrating across the Design Build Test cycle. Nat. Prod. Rep. **33**(8), 925–932 (2016). https://doi.org/10.1039/C6NP00018E
3. Carbonell, P., Jervis, A.J., Robinson, C.J., Yan, C., Dunstan, M., Swainston, N., Vinaixa, M., Hollywood, K.A., Currin, A., Rattray, N.J.W., Taylor, S., Spiess, R., Sung, R., Williams, A.R., Fellows, D., Stanford, N.J.,

1

Mulherin, P., Le Feuvre, R., Barran, P., Goodacre, R., Turner, N.J., Goble, C., Chen, G.G., Kell, D.B., Micklefield, J., Breitling, R., Takano, E., Faulon, J.L., Scrutton, N.S.: An automated Design-Build-Test-Learn pipeline for enhanced microbial production of fine chemicals. Commun. Biol. **1**(1), 66 (2018). https://doi.org/10.1038/s42003-018-0076-9

4. Chao, R., Yuan, Y., Zhao, H.: Recent advances in DNA assembly technologies. FEMS Yeast Res. **15**(1), n/a–n/a (2014). https://doi.org/10.1111/1567-1364.12171

5. Clomburg, J.M., Crumbley, A.M., Gonzalez, R.: Industrial biomanufacturing: the future of chemical production. Science **355**(6320), aag0804 (2017). https://doi.org/10.1126/science.aag0804

6. Dudley, Q.M., Karim, A.S., Jewett, M.C.: Cell-free metabolic engineering: biomanufacturing beyond the cell. Biotechnol. J. **10**(1), 69–82 (2015). https://doi.org/10.1002/biot.201400330

7. Lara, A.R., Palomares, L.A., Ramírez, O.T.: Scale-down: simulating large-scale cultures in the laboratory. In: Industrial Biotechnology, pp. 55–79. Wiley-VCH Verlag GmbH & Co. KGaA, Weinheim (2016). https://doi.org/10.1002/9783527807833.ch2

8. Nielsen, J., Keasling, J.D.: Engineering cellular metabolism. Cell **164**(6), 1185–1197 (2016). https://doi.org/10.1016/j.cell.2016.02.004

9. Polizzi, K., Stanbrough L, Heap, J.: A new lease of life, Understanding the risks of synthetic biology. Tech. rep., Lloyd's of London (2018). https://www.lloyds.com/~/media/files/news-and-insight/risk-insight/2018/a-new-lease-of-life/emerging-risk-report-2018---a-new-lease-of-life.pdf

10. Synthace: Computer Aided Biology. Tech. rep., Synthace (2018). https://synthace.com/computer-aided-biology-whitepaper

11. Wintle, B.C., Boehm, C.R., Rhodes, C., Molloy, J.C., Millett, P., Adam, L., Breitling, R., Carlson, R., Casagrande, R., Dando, M., Doubleday, R., Drexler, E., Edwards, B., Ellis, T., Evans, N.G., Hammond, R., Haseloff, J., Kahl, L., Kuiken, T., Lichman, B.R., Matthewman, C.A., Napier, J.A., ÓhÉigeartaigh, S.S., Patron, N.J., Perello, E., Shapira, P., Tait, J., Takano, E., Sutherland, W.J.: A transatlantic perspective on 20 emerging issues in biological engineering. eLife **6** (2017). https://doi.org/10.7554/eLife.30247

Further Reading

Some good introductions to **biochemistry and molecular cell biology**[1]:

Berg, J.M., Tymoczko, J.L., Stryer, L.: Biochemistry. Freeman (2011)

Lodish, H., Berk, A., Zipursky, S.L., Matsudaira, P., Baltimore, D., James Darnell, J.: Molecular Cell Biology. Freeman (2013).

Nelson, D.L., Cox, M.M.: Lehninger Principles of Biochemistry. Freeman (2013)

Alberts, B., Johnson, A., Lewis, J., Raff, M., Roberts, K., Walter P.: Molecular Biology of the Cell. Garland Science (2014)

Some introductions to **biotechnology** recommended for this textbook:

Villadsen, J., Nielsen, J., Lidén, G.: Bioreaction Engineering Principles. Springer (2011)

Villadsen, J., Lee, S.Y., Nielsen, J. (ed.): Fundamental Bioengineering. John Wiley (2016)

Renneberg, R., Berkling, V., Loroch, V.: Biotechnology for Beginners. Springer (2016)

In order to know more about **synthetic biology**:

Smolke, C., Lee, S.Y., Nielsen, J. (ed.): Synthetic Biology: Parts, Devices and Applications. John Wiley (2016)

Nesbeth, D.N.: Synthetic Biology Handbook. CRC Press (2016)

Baldwin, G., Bayer, T., Dickinson, R., Ellis, T., Freemont, P.S., Kitney, R.I., Polizzi, K., Stan, G.B.: Synthetic Biology – A Primer. World Scientific (2015)

---

1   Versions of some of these textbooks are freely available online at NCBI (▶ https://www.ncbi.nlm.nih.gov/books/).

Genome-Scale Modeling

© Springer Nature Switzerland AG 2019
P. Carbonell, *Metabolic Pathway Design*, Learning Materials in Biosciences,
https://doi.org/10.1007/978-3-030-29865-4_2

2

What You Will Learn in This Chapter
Before designing a new metabolic pathway, we need to learn how to model the cell behavior, how to relate its genome to its phenotype and how to simulate multiple growth conditions. In this chapter, we will have a look at some of the genome-scale models that the systems biology community has developed in the past few decades and will assess how these models can assist us in order to understand the cell, model the metabolic network and pathways and even predict the evolution of cell cultures. We will learn how to simulate the equilibrium state of the metabolite fluxes in the cell and how to evaluate cell capabilities depending on their context.

2.1 Systems Biology Models

Several years ago, cell cultures were basically modeled based on their macroscopic properties. Such models were useful in order to predict growth and mass transfer balances and are still an important part of the studies in biochemical engineering in order to model fermentation and transfer processes. One of these earlier models was the Monod model [3]. Even if these macroscopic models remain still useful in order to optimize bioprocesses [4], they do not greatly help us at the first stages of the metabolic pathway design. As metabolic pathway designers, the goal is to develop an industrial strain with the ability of producing some desired chemical. What we would like is to have a model that assists us in order to generate and test hypothesis about the effects of introducing some genes in a chassis organism. For that purpose, we need a model that looks at the balance of the chemical reactions occurring in the cell. Such endeavor might have seemed out of reach some decades ago but thanks to several high-throughput techniques developed in the last 20 years such as those for sequencing, proteomics, and metabolomics, we are able today to enjoy a clearer picture of the metabolism of the cell and the interplay of their main associated enzymes, reactions and metabolites. The reconstruction of **genome-scale models** is, along with genome sequencing and editing, one of the most successful community-wide efforts in providing new capabilities for biological engineering.

Underpinned by current high-throughput technologies, balancing the thousands of metabolic reactions and fluxes in a cell has shown to be an approachable problem made possible by our current computational capabilities and metabolic knowledge. Modeling dynamics and regulation, in turn, are problems whose complete resolution are still lagging behind. Solving the set of reachable equilibrium states in a metabolic network is relatively easy. Inconsistencies would mainly appear due to some missing reactions in the model or because some of them were assumed to be active while there were actually turned off by the cell. Solving dynamics and introducing regulation on the other hand is a more complex issue that we will cover in the next chapter.

2.2 Model Reconstruction from Omics to Big Data

Metabolic network reconstruction of a cell is a process that requires large efforts and resources, generally spanning many years of research by a trans-national consortium of teams. The goal is to identify every piece of the network that defines the metabolism of an organism, i.e., every biochemical reaction that is present at the highest level of detail.

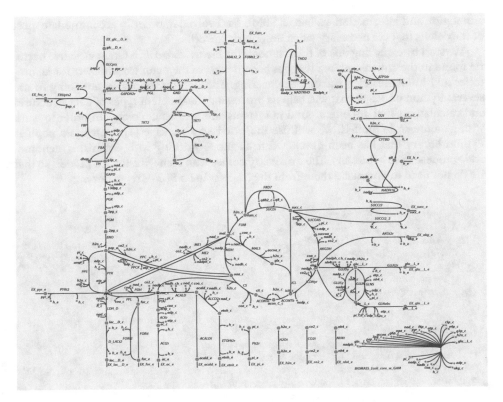

Fig. 2.1 Graphical representation of the *E. coli* core metabolism model

Network reconstruction is generally achieved one reaction at a time from the accumulated knowledge about the cell and the data available from genomics, transcriptomics, proteomics, metabolomics, etc. Every time a new network is satisfactorily reconstructed for some cell or organism, its publication hits the news headlines and is cheered up by the community. Everyone renders tribute to such achievement and teams world-wide start adopting this new model on their daily research.

A metabolic network is built based on its reactions. The network tells the information that links the metabolites and the reactions through their stoichiometry. Such relationships can be graphically represented in order to facilitate a quick overview of the different interactions (see ⬛ Fig. 2.1 for a graphical representation of the *Escherichia coli* core metabolism using the `Escher`[1] visualization system), but more importantly they can be represented using a mathematical representation. Such representation that links metabolites and the enzymatic reactions is the **stoichiometric matrix**. We will come back later in this chapter into a more detailed discussion about the properties of stoichiometric matrices.

Reconstructed metabolic models are generally deposited at the `BioModels`[2] database. The Systems Biology community has developed a standard representation for biological models that is called SBML (Systems Biology Markup Language). It allows better

---

1 ▶ https://escher.github.io/
2 ▶ https://www.ebi.ac.uk/biomodels-main/

2

annotation and information exchange. SBML has evolved in order to accommodate most of the details that are necessary when describing the cell.

We start by exploring one of these models for *Escherichia coli*, a Gram-negative bacteria found in the gut-microbiota that has become one of the main workhorsers for laboratories and industries in metabolic engineering. Through our book, we are going to use several Python packages that will help us processing the different types of data structures and calculations that are often found in systems and synthetic biology. In order to work with genome-scale models, we will use the `cobrapy` package [1], which is a popular Python library that has been developed to facilitate working with organisms' genome-scale models[3] (See ► Box 2.1). The `cobrapy` comes with some pre-loaded models so that we do not need to download them from the `BioModels` SBML repository.

Box 2.1

There exist numerous software tools that allow working with genome-scale metabolic models, including network reconstruction, analysis, and simulations [2]. Some work as stand-alone tools that can be installed in the computer, some are library packages in different languages such as MATLAB or Python, others offer web-based services. We will use some of these tools for different purposes in this textbook. A non-exhaustive list of some of the most important packages includes:

- `CellDesigner` (► http://http://www.celldesigner.org) is a structured diagram editor for drawing gene-regulatory and biochemical networks.
- `COPASI` (► http://copasi.org/) is a software application for simulation and analysis of biochemical networks and their dynamics.
- `OptFlux` (► http://www.optflux.org/) is a java-based software for constraint-based analysis that can perform many useful calculations in a friendly manner.
- `cobrapy` (► https://opencobra.github.io/) is a python package that provides a simple interface to metabolic constraint-based reconstruction and analysis.
- `Cameo` (► http://cameo.bio/) is a high-level python library developed to aid the strain design process in metabolic engineering projects.
- `Tellurium` (► http://tellurium.analogmachine.org/) is a Python environment for reproducible dynamical modeling of biological networks.

As described in Appendix A, you need to set up an environment in your computer using `Anaconda`[4] in order to install the required libraries. The process is relatively easy and you should be able to have the computer quickly configured so that you can start working with the examples given here. Secondly, you need some integrated development environment. As described in Appendix A, `Eclipse` is a classical environment that is popular among software developers and is a possible solution in order to start working with the examples described here. However, our main goal as metabolic pathway designers is not to develop new software but to perform **data analysis, simulation, modeling and design**. Therefore other environments that are more oriented towards scientific computing are preferred, like `Jupyter notebook` based on `IPython` or `Spyder`, a `MATLAB`-like open source environment for scientific computing in Python.

---

3   Full documentation and details about the `cobrapy` package are available at ► https://opencobra. github.io/

4   ► https://www.anaconda.com/

◻ **Table 2.1** Summary of the *E. coli* genome-scale model iJO1366

Type	Description
Name	iJO1366
Number of metabolites	1805
Number of reactions	2583
Objective expression	`-1.0*Ec_biomass_iJO1366_core_53p95M_reverse_e94eb`
	`+ 1.0*Ec_biomass_iJO1366_core_53p95M`
Compartments	Cytoplasm, periplasm, extracellular

Let's get our hands now on the first example. First, load the *E. coli* model that comes as example in `cobrapy`, shown in Code 2.1. Calling this code within a `Jupyter` notebook or `IPython` console will print the summary information of the model like in ◻ Table 2.1.

■■ **Code 2.1 Load a model of** *E. coli* **and print the information shown in** ◻ Table 2.1.

```
import cobra.test
model = cobra.test.create_test_model("ecoli")
display(model)
```

In this example, a model of *E. coli* called `iJO1366` (which was actually published in 2011 [5]) was loaded into our system containing 1805 metabolites and 2583 reactions (including transport reactions). These numbers give us already an idea of the complexity of such models. In the present model for a prokaryote cell like *E. coli* the number of cellular compartments is just three: cytoplasm, periplasm and extracellular, but this can become even more complicated for eukaryote cells with multiple compartments and organelles such as for the models proposed for the bakers yeast *Saccharomyces cerevisiae* (extracellular space, cytosol, mitochondrion, peroxisome, nucleus, endoplasmic reticulum, Golgi apparatus, or vacuole). Moreover, metabolic models for multicellular organisms like plants or human are multi-tissue and will therefore contain several tissue-specific models.[5] In genome-scale models, metabolites present in each compartment are typically considered as different chemical species, a convention that helps for instance in order to estimate fluxes between each cell's compartment and between the cell and the growth media. The objective expression consists of some combination of reactions that are assumed to model the cell behavior and will be discussed in the next ▶ Sect. 2.3.

Definitions and annotations for each metabolite, reaction or gene can be individually investigated. Any metabolite is identified by some identifier that corresponds to a short-

---

5 Details about available genome-scale models can be found in the additional references given at the end of this chapter.

2

hand representation of its name as well as its localization separated by an underscore symbol. For instance, the following code will generate a list of metabolites based on their id and common name:

■■ Code 2.2 Generate the metabolites list.

```
metabolites = [(x.id,x.name) for x in model.metabolites]
```

Each metabolite in the model is a Python class containing several fields with information. This is useful in order to search for specific classes of metabolites, like in using the molecular formula in order to search for some metabolite in the model. Consider the case that we are looking for D-glucose, its molecular formula is $C_6H_{12}O_6$. Code 2.3 will do the search for all the metabolites in the model having such molecular formula:

■■ Code 2.3 Print *E. coli* metabolites matching the molecular formula for glucose.

```
for x in model.metabolites:
    if x.formula == 'C6H12O6':
        print((x.id, x.name))
```

The result of running the metabolite search is shown in ◻ Table 2.2. As expected, the result contains D-glucose but also many isomers, which is something that occurs especially when dealing with sugars because of their large number of isomers.[6]

2.3 Model Simulation Through Constraint-Based Approaches

Genome-scale models can contain multiple information, the most basic is the stoichiometric relationships between the metabolites in the cell. As far as the reaction is feasible in the cell, most likely because of the presence of an enzyme that catalyzes it, chemical species must obey the stoichiometric equation. However, stoichiometric relationships do not provide enough information in order to know individual concentrations of metabolites because they are focused on the balance between reactants. Knowing the stoichiometry of the reactions, nevertheless, provides a basic information in order to understand the chemical capabilities and constraints of cell's metabolism. For instance the reaction labeled as PGI (see Code 2.4) performs a glucose-6-phosphate isomerase conversion that converts D-glucose 6-phosphate into D-fructose 6-phosphate, as summarized in ◻ Table 2.3.

---

6   The problem of isomer ambiguity is very common in mass spectroscopy where knowing the molecular formula and therefore the mass of a compound is not enough in order to unequivocally identify the actual chemical species.

■ **Table 2.2** Matched metabolites in *E. coli* with the glucose molecular formula

Label	Name
all__D_c	D-Allose
fru_c	D-Fructose
gal_c	D-Galactose
glc__D_c	D-Glucose
inost_c	myo-Inositol
man_c	D-Mannose
all__D_e	D-Allose
fru_e	D-Fructose
gal_e	D-Galactose
gal__bD_e	beta D-Galactose
glc__D_e	D-Glucose
inost_e	myo-Inositol
man_e	D-Mannose
all__D_p	D-Allose
fru_p	D-Fructose
gal_p	D-Galactose
gal__bD_p	beta D-Galactose
glc__D_p	D-Glucose
inost_p	myo-Inositol
man_p	D-Mannose

■ **Table 2.3** Summary of the PGI reaction in the *E. coli* model

Type	Description
Reaction identifier	PGI
Name	glucose-6-phosphate isomerase
Stoichiometry	g6p_c <=>f6p_c
	D-Glucose 6-phosphate <=> *D*-Fructose 6-phosphate
GPR	b4025
Lower bound	− 1000.0
Upper bound	1000.0

■■ **Code 2.4 Print the `PGI` reaction information of the *E. coli* model.**

```
pgi = model.reactions.get_by_id("PGI")
display(pgi)
```

In ◻ Table 2.3 we can see important information about the reaction. First, we get the definition, i.e., enzyme name and its stoichiometry. That is, the relative amounts of each reactant based on the law of conservation of mass. In addition, the GPR or gene-protein-reaction relationships tell us which gene or set of genes encode the enzyme(s) catalyzing the reaction in *E. coli*. In this case, the enzyme is encoded by gene `b4025` but other more complex cases involving multiple isoforms and subunits are possible. Such information would be helpful in case that we want to associate information from gene expression data to the metabolism of the cell. Moreover, the reaction is constrained with upper and lower bounds. Those bounds define what are the maximum and minimum limits in the flux associated with the reaction. For instance, in the example in ◻ Table 2.3, the flux will be given by the rate of interconversion of `g6p_c` into `f6p_c`, which will be given by the time change of the concentrations of each chemical species, i.e., $x_{g6p}(t)$, $x_{f6p}(t)$ respectively:

$$v_{PGI} = \frac{d\,x_{g6p}}{dt} = -\frac{d\,x_{f6p}}{dt}. \tag{2.1}$$

In general, a metabolite can participate in multiple reactions. Therefore, the total rate of consumption (or production) of the metabolite will be given by the sum of the fluxes of the reactions consuming (producing) the metabolite. As an example, we can look for all reactions in the model containing glucose (represented by the identifier `glc__D_c` in the model) as shown in Code 2.5, which will output the list shown in Code 2.6.

■■ **Code 2.5 List all reactions in the *E. coli* involving glucose.**

```
glucose = model.metabolites.get_by_id('glc__D_c')
for r in model.reactions:
    if glucose in r.reactants:
        print( (r.id, r.name, r.build_reaction_string()) )
```

■■ **Code 2.6 Output list of reactions involving glucose generated by Code 2.5.**

```
('GLCATr ', 'D-glucose_O-acetyltransferase', 'accoa_c_+_glc__D_c_
    <=>_acglc__D_c_+_coa_c')
('HEX1 ', 'hexokinase_(D-glucose:ATP )', 'atp_c_+_glc__D_c_-->_
    adp_c_+_g6p_c_+_h_c')
('XYLI2 ', 'xylose _isomerase', 'glc__D_c_<=>_fru_c')
```

When dealing with the full set of reactions in the model, we will generally use a matrix representation. Let's consider the following two reactions:

$$R_1 : \quad C_1 + C_2 \quad \rightleftharpoons \quad C_3 + C_4, \tag{2.2}$$
$$R_2 : \qquad C_3 \quad \rightarrow \quad C_5,$$

such reactions can be represented using the stoichiometric matrix **S** containing the stoichiometric coefficients:

$$\mathbf{S} = \begin{array}{cc} R_1 & R_2 \end{array} \atop \begin{pmatrix} -1 & 0 \\ -1 & 0 \\ 1 & -1 \\ 1 & 0 \\ 0 & 1 \end{pmatrix} \begin{array}{c} C_1 \\ C_2 \\ C_3 \\ C_4 \\ C_5 \end{array}. \tag{2.3}$$

In a similar fashion, we can combine together in a stoichiometric matrix **S** the full set of reactions and metabolites that are present in the model. Because of the law of conservation of mass, the change in time of each individual concentration $x_i(t)$ has to equal the sum of reaction fluxes weighted by their corresponding stoichiometric coefficients:

$$\dot{x} = \frac{d\mathbf{x}}{dt} = \mathbf{Sv}. \tag{2.4}$$

Now, let's assume that the cell is in equilibrium (see ▶ Box 2.2) and therefore the concentrations are constant:

$$\frac{d\mathbf{x}}{dt} = \mathbf{0}. \tag{2.5}$$

Box 2.2   Cell's Equilibrium
Living organisms are never in equilibrium as they are constantly adapting their internal metabolism to environmental changes. However, we can assume that under certain conditions a cell population will behave as if it was in equilibrium. Equilibrium is understood here as "everything that goes in should go out". In other words, no accumulation or depletion of chemicals. Since a cell needs to consume some nutrients in order to maintain its equilibrium, a constant influx of precursor nutrients into the cell's culture and a constant downstream collection of waste is always running. Such continuous exchange conditions are reproduced in bioreactors by systems like the chemostat.

Genome-scale models represent such input/output influx either by using an unbalanced reaction like $\cdot \rightarrow glc$ and $co2 \rightarrow \cdot$, or by defining some boundary compartment that would be considered as the input/output exchange with the environment.

Microorganism populations like those from *E. coli* are controlled in industrial environments in order to go through different phases like growth and production phases, each one showing different metabolic equilibrium. Genome-scale models express fluxes in relative terms to the biomass, often in units such as $mmol \times h^{-1} \times gDW^{-1}$, where gDW is gram dry weight, a proxy for biomass $N(t)$ (number of cells). Therefore, equilibrium is understood as equilibrium per unit of biomass, which is something that we assume that occurs during the exponential phase.

During the exponential phase, cells are growing at a rate that is proportional to its population and verifies the following relationship:

$$\frac{dN}{dt} = \mu N(t) \tag{2.6}$$

being μ the effective growth rate or doubling time. Since the change in biomass is proportional to the number of cells and units are normalized by the biomass, we can assume that the stoichiometric equilibrium of one cell will be equivalent to the macroscopic equilibrium of the total population.

In such case Eq. 2.4 will be simplified as follows:

$$\mathbf{0} = \mathbf{Sv}, \tag{2.7}$$

which defines a set of linear constraints between the fluxes in the cell. Now, trying to manually solve such highly-dimensional equation is not possible in practice (remember from ◻ Table 2.1 that the model of *E. coli* contains 1805 metabolites and 2583 reactions, which implies a stoichiometric matrix of dimension 1805×2583!), but it can be easily solved using some scientific computing software such as MATLAB, numpy in Python or R. As you can imagine, most of the elements in the stoichiometric matrix **S** contain zeros since each reaction and its associated columns typically involve only a small number of metabolites. Highly-efficient optimization algorithms exist that can be applied in order to solve sparse equations like the one in Eq. 2.7.

Another important consequence of the highly-dimensional nature of Eq. 2.7 is that solving it will lead to infinite solutions, i.e., there exist many possible combinations of fluxes **v** that can verify the equation. In fact, rather than considering single solutions, we need to consider a region or subspace of feasible fluxes solutions. Each of those solutions v_i can be interpreted as a state of the cell. The question is therefore, even if we were able to determine the full set of solutions, how do we know which state among the full set is the one that better represents the actual state of the cell? In order to answer this question, we need to define the **context of the cell**:

- **Flux constraints**: as we have previously seen in ◻ Table 2.3, each flux v_i is associated with upper v_i^u and lower v_i^l bounds. Such bounds can be used not only to define the available precursors in the growth media but also to delimit the actual capabilities of each enzymatic reaction in the forward (positive) and reverse (negative) directions;

- **Objective function**: besides the model constraints, we can define some objective function that tries to reproduce the actual preferred state of the cell. Objective functions are generally expressed as a linear combination of the fluxes: $J = \mathbf{x}^T \mathbf{v}$. One objective function that is typically chosen is the biomass function, i.e., to assume that a cell population will always try to maximize its biomass. Under such assumption, a biomass function can be approximated by looking at the main composition of the dry extract of the cell.

Therefore, the solution is given by the following optimization problem:

$$\max \quad J = \mathbf{x}^T \mathbf{v}$$

subject to:

$$0 = \mathbf{Sv}$$

$$\mathbf{v}^l \leq \mathbf{v} \leq \mathbf{v}^u.$$

(2.7)

For instance, the objective function of our *E. coli* model, shown in Code 2.7, contains a biomass core function (also a reverse term that can be ignored here as it is set to zero):

■■ **Code 2.7 Print the objective function of the** *E. coli* **model.**

```
model.objective.expression
 -1.0* Ec_biomass_iJO1366_core_53p95M_reverse_e94eb + 1.0*
    Ec_biomass_iJO1366_core_53p95M
print(model.reactions.getby_id('Ec_biomass_iJO1366_core_53p95M')
    )
```

If we keep both the default bounds of the fluxes and the biomass objective of the model in the cobrapy *E. coli* model, solving Eq. 2.7 can be performed as shown in Code 2.8. Summary fluxes in this example will print one of the possible flux distribution solutions. However, flux solutions might belong to a full region, as previously discussed. In the next section, we will describe a method that can be used with cobrapy to determine the boundaries of the full set of flux solutions.

■■ **Code 2.8 Print solution fluxes.**

```
model.optimize()
model.summary()
IN FLUXES            OUT FLUXES              OBJECTIVES

o2_e       17.6      h2o_e      45.6         Ec_biomass_i...  0.982
nh4_e      10.6      co2_e      19.7
glc__D_e   10        h_e        9.03
pi_e       0.948     mththf_c   0.00044
so4_e      0.248     5drib_c    0.000221
k_e        0.192     4crsol_c   0.000219
fe2_e      0.0158    amob_c     1.96e-06
mg2_e      0.00852   meoh_e     1.96e-06
ca2_e      0.00511
cl_e       0.00511
cu2_e      0.000697
mn2_e      0.000679
zn2_e      0.000335
ni2_e      0.000317
mobd_e     0.000127
cobalt2_e  2.46e-05
```

2.4 Advanced Applications of Flux Analysis

Metabolic flux analysis is one of the most important achievements in systems biology. There are many types of analysis that can be performed based on this elegant technique. For instance, flux balance analysis (FBA) would often lead not to a unique solution but to a whole range of alternative solutions, all of them optimal in the sense that they optimize the objective function and simultaneously verify all the imposed flux constraints. A simple approach in order to inspect the set of optimal fluxes is to determine the range of allowed fluxes in the solutions. A reaction that has a short range of valid fluxes might indicate an important reaction that is less flexible or has less *elasticity*. Such type of analysis is known as **flux variability analysis** (FVA) and consists on finding the allowed range of each flux at the optimal solution. FVA is implemented in cobrapy. Code 2.9 will compute the flux range for the xylose isomerase reaction shown in Code 2.10:

■■ Code 2.9 Perform flux variability analysis for xylose isomerase.

```
from cobra.flux_analysis import flux_variability_analysis
fv = flux_variability_analysis( model, [model.reactions.getby_id
    ('XYLI2')])
display( fv )
```

■■ Code 2.10 Output of the flux variability analysis computed in Code 2.9.

	maximum	minimum
XYLI2	7.597609e − 12	− 2.890889e− 12

FBA is a powerful approach to estimating the equilibrium fluxes or steady states of the cell under some given conditions. However, cell conditions can change over time. **Dynamic FBA** is an FBA extension that consists on re-running FBA under changing conditions, for instance addition or depletion of some key growth media components, change in pH conditions, etc. This is especially useful in cases where some key metabolite concentrations are continuously measured so that the FBA calculation can be re-run for the new conditions in order to estimate the new distribution of fluxes. Dynamic FBA performs a dynamic simulation that calculates multiple pseudo-equilibrium states. Cells are nevertheless in continuous change and sometimes it will be useful to also look at the transient behavior of the dynamic response of key elements of the metabolic engineering of the pathway. In the next ▶ Chap. 3, we will discuss how to run simulations and analyze the dynamic response of pathways.

Many other types of flux analyses are possible [2]. We had here a first introduction to the possibilities of genome-scale models and metabolic flux analysis. Nowadays there is a vibrant community of researchers and entrepreneurs contributing to the field of metabolic flux analysis. We will come back later in this book in order to learn more about genome-scale models in metabolic pathway engineering once we have visited other concepts related to synthetic biology so that we can modify the initial model by introducing genetic circuits.

Take Home Message

- Genome-scale models provide a detailed account of the metabolic reactions taking place in the cell.
- Genome-scale models have been reconstructed for several organisms by the systems biology community.
- Constraint-based flux analysis calculates the optimal distribution of fluxes in the model based on the growth conditions and some objective function.
- Objective functions can represent biomass formation as well as other objectives that the cell tries to maximize.
- Standard representation of the models can be achieved by using the Systems Biology Mark-up Language (SBML).

2.5 Problems

? **2.1** Look for amino acids in a genome-scale model.

✓ The list of 20 proteinogenic amino acids (the ones that are naturally incorporated into proteins during translation) are listed in ◼ Table 2.4 along with their chemical formula and molecular weight. Following our previous example in Code 2.3, look for metabolites in the *E. coli* model with same chemical formula as the 20 amino acids. How many of them contain isomers in the model?

? **2.2 Shared reactions in two genome-scale models.**

✓ In addition to the *E. coli* model, the `cobrapy` package comes with a bundled model for *Salmonella* under the name `"salmonella"`. Both models contain approximately the same number of metabolites. Calculate how many metabolites they have in common. What about the common reactions?

? **2.3 Metabolites in the biomass function.**

✓ How many metabolites contain the biomass function in the *E. coli* model? Which metabolites have the higher stoichiometric coefficients?

? **2.4** Change objective function.

✓ Code 2.8 calculated the solutions to Eq. 2.7 for the default biomass function. In some cases, we might be interested in solving the equation for other objectives, which can be changed in the model by redefining `model.objective`. Calculate the solutions for the *E. coli* model with the following objectives:
1. ATP consumption;
2. Redox exchange;
3. Sum of amino acid fluxes.

Hint: look for fluxes involving each of the terms and redefine the objective function based on those fluxes.

2

⬛ Table 2.4 Amino acids list

Amino acid	Chemical formula	Molecular weight [g/mol]
Isoleucine	C6H13NO2	131.1736
Leucine	C6H13NO2	131.1736
Lysine	C6H14N2O2	146.1882
Methionine	C5H11NO2S	149.2124
Phenylalanine	C9H11NO2	165.1900
Threonine	C4H9NO3	119.1197
Tryptophan	C11H12N2O2	204.2262
Valine	C5H11NO2	117.1469
Arginine	C6H14N4O2	174.2017
Histidine	C6H9N3O2	155.1552
Alanine	C3H7NO2	89.0935
Asparagine	C4H8N2O3	132.1184
Aspartate	C4H7NO4	133.1032
Cysteine	C3H7NO2S	121.1590
Glutamate	C5H9NO4	147.1299
Glutamine	C5H10N2O3	146.1451
Glycine	C2H5NO2	75.0669
Proline	C5H9NO2	115.1310
Serine	C3H7NO3	105.0930
Tyrosine	C9H11NO3	181.1894

❓ 2.5 Simulate aerobic and anaerobic conditions.

✅ Simulate the *E. coli* model and calculate the flux solutions in aerobic and anaerobic conditions. *Hint:* change oxygen availability in order to simulate both conditions.

❓ 2.6 Carbon source ranges.

✅ Considering glucose and glycerol as carbon sources of the *E. coli* model, perform a flux variability analysis like the one shown in Code 2.9 and calculate the allowed ranges for glucose and glycerol. *Hint:* find the fluxes producing both feedstock precursors (glucose and glycerol).

? **2.7** Simulate multiple hosts.

✓ Go to the `BioModels` database.[7] Download a genome-scale model for some industrial chassis like: (a) yeast (*Saccharomyces cerevisiae*), (b) *Bacillus subtilis*, (c) *Pichia pastoris*. Simulate the models and compare their maximum achievable biomass.

? **2.8** Simulate anaerobic conditions in yeast.

✓ Take the yeast (*Saccharomyces cerevisiae*) model from the previous exercise and simulate it under anaerobic conditions. What is the maximum predicted biomass of the anaerobic yeast? *Hint:* set oxygen incoming flux to zero in order to simulate anaerobic conditions.

? **2.9** Simulate gene essentiality.

✓ Simulate the *E. coli* model under deletion of proteinogenic amino acids producing genes in ◻ Table 2.4. Are these genes essential, i.e., the cell does not grow upon their deletion? *Hint:* look for the fluxes producing each amino acid and delete them from the model in each simulation.

? **2.10** Simulate synthetic lethality.

✓ Simulate now the *E. coli* model for synthetic lethality under simultaneous deletion of two proteinogenic amino acids producing genes in ◻ Table 2.4. In which cases of double deletion the cell is able to survive?

References

1. Ebrahim, A., Lerman, J.A.J., Palsson, B.O., Hyduke, D.R.: COBRApy: COnstraints-based reconstruction and analysis for python. BMC Syst Biol. **7**(1), 74 (2013). https://doi.org/10.1186/1752-0509-7-74
2. Hamilton, J.J., Reed, J.L.: Software platforms to facilitate reconstructing genome-scale metabolic networks. Environ. Microbiol. **16**(1), 49–59 (2014). https://doi.org/10.1111/1462-2920.12312
3. Klipp, E., Liebermeister, W., Wierling, C., Kowald, A., Lehrach, H., Herwig, R.: Systems Biology: A Textbook. Wiley-Blackwell, Chichester (2009). https://doi.org/10.3797/scipharm
4. Nielsen, J., Villadsen, J., Lidén, G.: Bioreaction Engineering: From Bioprocess Design to Systems Biology. In: Bioreaction Engineering Principles, pp. 1–8. Springer US, Boston (2003). https://doi.org/10.1007/978-1-4615-0767-3_1
5. Orth, J.D., Conrad, T.M., Na, J., Lerman, J.A., Nam, H., Feist, A.M., Palsson, B.O.: A comprehensive genome-scale reconstruction of Escherichia coli metabolism-2011. Mol. Syst. Biol. **7**, 53 (2011). https://doi.org/10.1038/msb.2011.65

Further Reading

The following provides a comprehensive review on **reconstruction of metabolic networks** as well as links to an up-to-date list of metabolic models with their properties (see its Supplementary Table S1):
Feist, A.M., Herrgård, M.J., Thiele, I., Reed, J.L., Palsson, B. Ø.: Reconstruction of biochemical networks in microorganisms. Nat. Rev. Microbiol. **7**(2), 129 (2009)

7 ▶ https://www.ebi.ac.uk/biomodels-main/

An excellent and comprehensive introduction to the field of **constraint-based analysis**:
Palsson, B.Ø.: Systems Biology. Cambridge University Press, Cambridge (2015)

Other more specialized sources on **metabolic network analysis** are the following:
Maranas, C.D., Zomorrodi, A.R.: Optimization Methods in Metabolic Networks. Wiley-Blackwell, New Jersey (2016)
Smolke, C.D. (ed.): The Metabolic Pathway Engineering Handbook. Tools and Applications. CRC Press/Taylor & Francis, Boca Raton (2010)

Pathway Modeling

© Springer Nature Switzerland AG 2019
P. Carbonell, *Metabolic Pathway Design*, Learning Materials in Biosciences,
https://doi.org/10.1007/978-3-030-29865-4_3

3

What You Will Learn in This Chapter

Introducing an engineered metabolic pathway into the cell not only alters its equilibrium but also modifies its dynamics. Modeling the transient behavior of an engineered pathway using kinetic models can provide valuable insights in order to identify what are the rate-limiting steps and how are they interrelated with the flux exchanges of endogenous precursors in the cell. Moreover, understanding the dynamics of the metabolic pathway is a first step on robustness analysis, providing clues about points of intervention in the pathway for transcriptional (promoters, terminators) or translational (ribosome binding site) control by means of selection of genetic parts. In this chapter, you will learn some basic models for the main genetic parts that are involved in an engineered metabolic pathway and the tools for simulating their dynamic response.

3.1 Pathway Steady-State and Dynamics

In previous ▶ Chap. 2 we learned how to model and simulate the cell state through genome-scale models. Because of the large number of metabolites and reactions that were involved in the model, simulations were restricted to determining the set of steady-state solutions verifying some given constraints. Even if some attempts have been carried out in order to simulate the whole-cell dynamics [4], the problem is complex and still largely remains as an open challenge because of the large number of assumptions and parameters that need to be defined and estimated.

Rather than trying to simulate the whole-cell we would like to focus on local modeling and simulation of an engineered metabolic pathway. Contrary to genome-scale models, engineered pathways only contain a relatively small number of metabolites and reactions, thus allowing for a more detailed simulation of their kinetics. An **engineered metabolic pathway** is a genetic circuit that expresses one or more enzymes in the cell in order to catalyze some biochemical reactions that will lead to the production of a target chemical. The basic backbone of an engineered pathway is a **promoter**, a **ribosome binding site** (RBS), a **coding sequence** (CDS) containing the enzyme DNA sequence and a **terminator**. A promoter is a DNA region that initiates transcription and it is used to tune binding to the RNA polymerase. There are several types of promoters showing different transcriptional activity: constitutive promoters are always active while inducible promoters are activated through the inducer. The RBS, in turn, is used in order to tune mRNA translation. It is responsible for the recruitment of a ribosome during the initiation of protein translation. The design of enzyme sequences will be discussed in ▶ Chap. 5 and the design of promoters and RBS will be discussed in ▶ Chap. 8. ◻ Figure 3.1 shows a standard representation of the pathway genetic construct using the

◻ **Fig. 3.1** Graphical representation of the DNA regions of a sequence containing a promoter, RBS, CDS and terminator

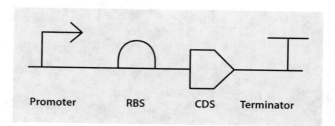

| Promoter | RBS | CDS | Terminator |

Synthetic Biology Open Language (SBOL) Visual specification, which will be discussed in more detailed later in ► Chap. 8.

From the point of view of the metabolic network, expressing an engineered metabolic pathway in the cell implies the inclusion of new reactions and metabolites into the model. Equation 2.4, which modeled the cell's metabolic network needs to be augmented with the new metabolite concentrations as follows:

$$\hat{\mathbf{x}} = \begin{bmatrix} \dot{\mathbf{x}} \\ \dot{\mathbf{x}}_p \end{bmatrix} = \begin{bmatrix} \mathbf{S} & \mathbf{S}_{xp} \\ \mathbf{S}_{px} & \mathbf{S}_p \end{bmatrix} \begin{bmatrix} \mathbf{v} \\ \mathbf{v}_p \end{bmatrix}. \tag{3.1}$$

Here, we would like to model and simulate **pathway dynamics** $\dot{\mathbf{x}}_p$, i.e., the time course of the metabolite concentrations that are involved in the pathway (Eq. 3.1). Note that this requires both the knowledge of the pathway kinetics as well as the time course evolution of the metabolite exchange between the pathway and the cell given by $\mathbf{S}_{px}\mathbf{v}$:

$$\dot{\mathbf{x}}_p = \begin{bmatrix} \mathbf{S}_{px} & \mathbf{S}_p \end{bmatrix} \begin{bmatrix} \mathbf{v} \\ \mathbf{v}_p \end{bmatrix}. \tag{3.2}$$

Similarly to the genome-scale analysis presented in ► Chap. 2, steady-state solutions of previous Eq. 3.2 (Eq. 3.3) can be computed through flux balance analysis:

$$0 = \begin{bmatrix} \mathbf{S}_{px} & \mathbf{S}_p \end{bmatrix} \begin{bmatrix} \mathbf{v} \\ \mathbf{v}_p \end{bmatrix}. \tag{3.3}$$

Pathway's steady-state based models are useful and easy to simulate at the beginning of our metabolic engineering project in order to assess theoretical capabilities of both the cell and the pathway. In ► Chap. 7, we will come back to this analysis and will apply it to an example using `cobrapy` to compute the steady-state solutions of an engineered flavonoid production pathway in *E. coli*.

Now, if we want to gain a better understanding of the interplay between the different elements of the pathway, we need a more detailed mechanistic model. Such detailed model will allow us to estimate metabolite concentrations, to analyze the effects of the different genetic parts in our circuit and to simulate their dynamics. The actual dynamics of a pathway is complex for multiple reasons:

- The pathway does not occur in isolation, but it competes with many others for the availability of the precursors;
- Fluctuations occur in protein concentrations, we might need to introduce a model for this uncertainty;
- Protein regulation, which introduces additional interactions and non-linearities;
- Environmental variations, here we will ignore the localization and crowding effects but they might introduce important effects in the behavior of the cell;
- The assumption that simulation of a single-cell model is valid in order to predict macroscale properties, which may be limited by dilution effects.

3

In order to perform our simulations, we will use in the rest of this chapter the COPASI software (a complex pathway simulator) [2].[1] A brief description of this tool is provided in ▶ Box 3.1. COPASI can be used for creation, modification, simulation and computational analysis of kinetic models. Many of the tasks that we will perform through this textbook can be carried out by using COPASI. COPASI can import and export SBML models and it is therefore compatible with the genome-scale models that were introduced in ▶ Chap. 2. Here, we will focus on the capabilities of COPASI for kinetic models, which are definitely one of its strengths. We will come back in later chapters into this software to perform other additional model analyses.

> **Box 3.1**
> COPASI is a modeling and simulation environment. The software can be installed in multiple environments from the COPASI website[2] and run as a stand-alone tool. Main features include model import using model standards such as SBML, model editing and simulation algorithms and multiple analysis tasks. An interesting feature of COPASI are their algorithms for model parameter estimation from experimental data. Even if the most popular implementation of COPASI is through the stand-alone tool, other versions exist as language interfaces to Java, Python, etc or through a web interface.

3.2   The Law of Mass Action

The mass action law is at the roots of biochemical kinetics. It states that a reaction rate is proportional to the probability of collision of the reactants. Such probability depends on the concentration of reactants (number of molecules of each species) raised to the number of molecular species that enter into the reaction (the stoichiometric coefficient). For instance the following reversible reaction with substrates S_1 and S_2 and product P_1 (1 molecule of S_1 and 2 molecules of S_2 react to form P_1 and 1 molecule of P_1 reacts to form 1 molecule of S_1 and 2 molecules of S_2):

$$S_1 + 2 S_2 \; \rightleftharpoons \; P_1, \tag{3.4}$$

will have the following net reaction rate v according to the law of mass action:

$$v \; = \; k_+ \, S_1 \cdot S_2^2 - k_- \, P_1, \tag{3.5}$$

where k_+ and k_- are the rate constants of the forward and backward reaction, respectively. These k_+ and k_- parameters are kinetic constants that are specific to each reaction.

In a general case with m_s substrates S_i and m_p products P_i, the law of mass action can be written as follows:

$$v = k_+ \prod_{i=1}^{m_s} S_i^{n_{si}} - k_- \prod_{i=1}^{m_p} P_i^{n_{pi}}, \tag{3.6}$$

---

1   An alternative to COPASI for running simulations is Tellurium ▶ http://tellurium.analogmachine. org/, which is a Python-based environment for modeling, simulation and analysis for systems and synthetic biology models.

2   ▶ http://copasi.org

where n_{si} and n_{pi} are the number of molecules of each substrate or product, respectively that participate in the reaction (molecularity).

When reaction 3.4 reaches the equilibrium, i.e., when the forward and backward rates become equal, according to Eq. 3.6, the ratio of product and substrate concentrations are related by the equilibrium constant K_{eq}:

$$K_{eq} = \frac{k_+}{k_-} = \frac{\prod_{i=1}^{m_p} P_i^{n_{pi}}}{\prod_{i=1}^{m_s} S_i^{n_{si}}}. \tag{3.7}$$

Concentrations outside the equilibrium will vary dynamically following the rate law for v given in Eq. 3.6:

$$\dot{S}_1 \quad = \quad \frac{dS_1}{dt} = -v, \tag{3.8a}$$

$$\dot{S}_2 \quad = \quad \frac{dS_2}{dt} = -2v, \tag{3.8b}$$

$$\dot{P}_1 \quad = \quad \frac{dP_1}{dt} = v. \tag{3.8c}$$

3.3 Modeling Enzyme Kinetics

Modeling enzyme kinetics has been for more than a century an integral part of biochemistry with a rich and colorful story. The Michaelis-Menten model is perhaps the most well-known and celebrated model. In its simplest form, it represents the rate of an irreversible mono-substrate reaction catalyzed by enzyme E and corresponds to:

$$E + S \underset{k_r}{\overset{k_f}{\rightleftharpoons}} ES \overset{k_{cat}}{\rightarrow} E + P, \tag{3.9}$$

where E stands for the enzyme, S for the substrate, P for the product and ES for the transition complex formed by the enzyme and the substrate. Equation 3.9 assumes that the role of enzyme E is to convert molecules of S into substrate P.

Applying the law of mass action (Eq. 3.6) to the previous reaction leads to the following set of ordinary differential equations (ODEs):

$$\frac{dS}{dt} \quad = \quad -k_f E \cdot S + k_r ES, \tag{3.10a}$$

$$\frac{dES}{dt} \quad = \quad k_f E \cdot S - \left(k_r + k_{cat}\right) ES, \tag{3.10b}$$

$$\frac{dE}{dt} = -k_f E \cdot S + \left(k_r + k_{cat}\right) ES,$$

(3.10c)

$$\frac{dP}{dt} = k_{cat} ES \cdot S.$$

(3.10d)

In the previous equation 3.10, the reaction rate v is the rate of product P formation and equals the consumption of substrate S:

$$v = \frac{dP}{dt} = -\frac{dS}{dt}.$$

(3.11)

The total amount of enzyme E_T is considered to remain constant, either in free form E or as part of the complex ES. By adding Eqs. 3.10b and 3.10c, we obtain the following constant relationship:

$$\frac{dES}{dt} + \frac{dS}{dt} = 0,$$

(3.12a)

$$E + ES = E_T = (constant).$$

(3.12b)

The set of equations 3.10 are generally not solved directly because of their complexity, but different assumptions are used in order to simplify the expression so that their parameters can be estimated. Generally it is considered that the reversible transient binding of the enzyme to the substrate is much faster than the conversion of the complex ES into the product P:

$$k_f, k_r \gg k_{cat}.$$

(3.13)

Moreover, the concentration of the ES complex is assumed to reach a *quasi steady-state* condition where it remains constant. This assumption requires that the initial substrate concentration S to be much larger than the enzyme concentration so that a point is reached where the enzyme is saturated. Under such assumption, we can rewrite Eq. 3.10c as follows:

$$ES = \frac{k_f E_T S}{k_f S + k_r + k_{cat}}.$$

(3.14)

We define the Michaelis-Menten constant:

$$K_M = \frac{k_r + k_{cat}}{k_f},$$

(3.15)

which equals to the substrate concentration that yields the half-maximal reaction rate. Substituting in Eq. 3.14:

$$ES = \frac{E_T S}{S + K_M},$$

(3.16)

which can be then substituted in Eq. 3.10d in order to obtain the expression of the reaction rate:

$$v = \frac{k_{cat} E_T S}{S + K_M}.$$ (3.17)

Finally, we introduce a new parameter called V_{max} defined as the maximal rate of product conversion that we can obtain if the enzyme is completely saturated. V_{max} is calculated as the product of the total concentration of enzyme E_T by k_{cat} or *turnover rate*:

$$V_{max} = k_{cat} E_T.$$ (3.18)

We arrive at a simplified expression for enzyme kinetics that is used often in practice:

$$v = \frac{E_T k_{cat} S}{K_M + S} = \frac{V_{max} S}{S + K_M}.$$ (3.19)

In Eq. 3.19, the reaction rate v depends only on a single variable, the substrate availability, and requires the determination of two parameters: V_{max} and K_M, in order to fit the equation. Interestingly, both parameters can be easily derived from an experimental assay by determining the dependence curve between reaction rates and substrate at different concentrations.

We start our simulation analysis by looking for kinetics data in public databases for the enzymes in the target pathway. In our example, we will consider a pathway construct that contains a single gene encoding a phenylalanine ammonia lyase (PAL) enzyme (EC number 4.3.1.24), whose reaction is shown in ◘ Fig. 3.2.

Kinetics parameters can be retrieved from databases such as SABIO-RK[3] or BRENDA.[4] A search in SABIO-RK for phenylalanine ammonia lyase will output multiple instances. For this example, we have selected the parameters from an experimental assay where a wildtype variant (a native protein without mutations) from parsley (*Petroselinum crispum*) was expressed in *E. coli*,[5] as shown in ◘ Table 3.1. After selecting in SABIO-RK this

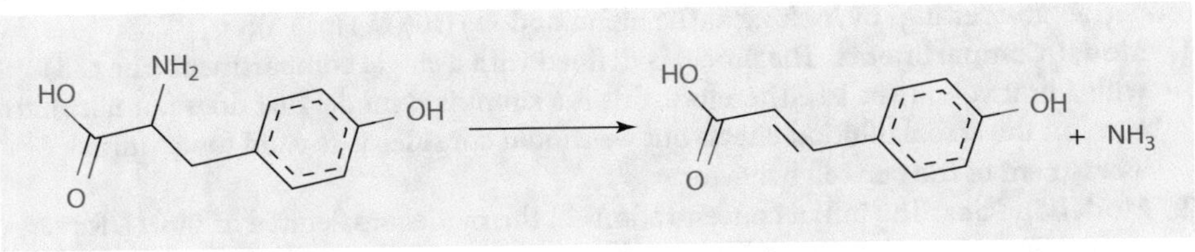

◘ **Fig. 3.2** Phenylalanine ammonia lyase reaction involving *L*-phenylalanine as substrate and trans-cinnamate and ammonia as products

---

3 ▶ http://sabio.h-its.org/
4 ▶ https://www.brenda-enzymes.org/
5 Entry ID: 22717.

3

■ **Table 3.1** Kinetic parameters of enzyme PAL from parsley source as downloaded from `SABIO-RK` database

Parameter	Value	Units
k_{cat}	13.5	s^{-1}
K_M (L-phenylalanine)	0.12	mM

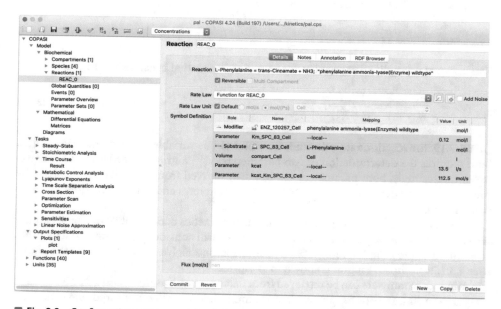

■ **Fig. 3.3** Configuration options and simulation parameters for the model in `COPASI`

particular case as an entry to export, the resulting model can be downloaded in the SBML exchange format (introduced in ▶ Chap. 2).

Once the model is imported into `COPASI`, we can inspect and modify their different properties and settings by looking at the menu options (see ■ Fig. 3.3):

1. **Model/Compartments**: The model is defined with a single compartment (the cell) with a unit volume of 1 L. Therefore, this is a simplified model that does not take into account the actual dilution effects but we should consider it as valid for an initial assessment of the per-cell behavior.

2. **Model/Species**: The initial concentrations in the model are defined as mol/L for each chemical species, i.e., of L-phenylalanine, trans-cinnamate, NH_3 and the PAL enzyme. Trans-cinnamate is an heterologous *E. coli* metabolite that will be only produced by the host once PAL will start being expressed. Therefore, we edit the model in order to change its initial concentration to 0.

3. **Model/Rate law**: If we inspect the reactions contained in the model we can see that the enzymatic reaction is defined through a reaction that is called `Function for REAC_0`. Such function is described under **Tasks/Functions**. In our case, the reaction rate law follows the Michaelis-Menten kinetics as defined by Eq. 3.19.

4. **Tasks/Time Course**: A simulation of the time course change of concentrations for the substrate and products will be run. Since there is a fast turnover rate (k_{cat}) of 13.5 s^{-1}, L-phenylalanine will be consumed relatively quickly once the transient substrate-enzyme complex has been established. Therefore, enzyme availability will have an important effect on the actual time scale of the transient concentration of the product. Here, we will consider an initial enzyme concentration 10 µM and will set up a simulation period for 2 h (7200 s).

5. **Output Specifications/Plots**: COPASI provides a plotting facility for the results. We can add a time course plot with two curves, one for the substrate L-phenylalanine and another one for the product trans-cinnamate.

As previously discussed, prior to run the simulation we need to fix the initial concentrations for the chemical species. ◻ Table 3.2 shows the chosen values for the initial concentrations of each molecule (they can be also given as number of molecules depending on the type of simulation). ◻ Figure 3.4 plots the transient concentrations of both the substrate L-phenylalanine and the product trans-cinnamate after running the simulation task in COPASI. We can see that it took around 1.5 hours for the substrate L-phenylalanine to be consumed and converted into the product trans-cinnamate.

3.4 Modeling Transcriptional and Translational Control

In our previous time-course analysis we assumed that the concentration of the enzyme was constant during the full period of the simulation. However, the actual concentrations of the enzymes will depend on the different types of regulation of gene expression associated with each enzyme. Native pathways in both prokaryotes and eukaryotes are generally regulated and therefore analyzing their dynamic behavior is more complex as it requires of detailed mechanistic models and time separation analysis. In an engineered metabolic pathway that is non-natively expressed in the host, gene expression would mainly depend on those genetic regulatory circuits that have been engineered as part of the design. Two types of control are generally used:

- **Transcriptional control** to tune the transcription activation and rate of the gene;
- **Translational control** to tune the translation rate of mRNA by the ribosomes.

Models exist that can predict both promoter and ribosome strengths, i.e., activity, from their DNA sequence through different approaches (mechanistic based on thermodynamics, machine learning, etc.). Some are available online like the RBS calculator [6].

◻ **Table 3.2** Initial concentrations for the kinetics simulation of the enzymatic reaction

Species	Concentration
(L-phenylalanine)	10 mM
NH$_3$	10 mM
PAL	10 µM
trans-cinnamate	0 mM

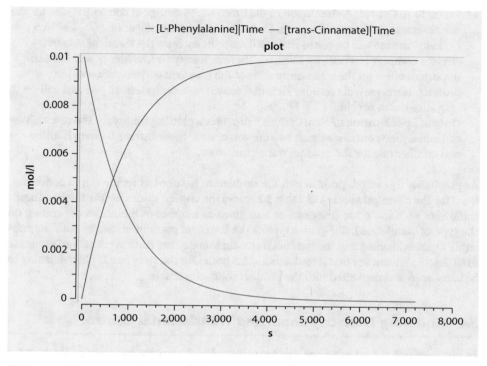

Fig. 3.4 Time course response of substrate *L*-phenylalanine (red curve) and product trans-cinnamate (blue curve)

Transcription regulation is controlled by promoters and terminators. A promoter defines the transcription starting and initiation points. Promoters often contain DNA motifs enabling the binding of specific transcription factors that act as repressors or as activators of RNA polymerase recruitment. Interestingly, these transcription factors can become activated or inactivated by inducers, which are specific chemical signals that switch the functionality of the transcription factor. Some promoters found in *E. coli* that are routinely used nowadays in pathway vectors in order to introduce controlled gene expression are:

- P_{tet} that can use transcription factors that act as either inducer or repressor in the presence of tetracycline or doxycycline;
- P_{lac}, P_{trc} or P_{lacO}, where the protein LacI targets the Lac operon with inducers lactose and IPTG (isopropyl-β-D-thiogalactoside).

Synthetic gene regulatory modules are often described with Hill kinetics [5] to model its switch-like behavior. A basic model of promoter activation of enzyme expression considers the inactivated promoter P, the activated promoter P^* and the inducer I. In the following reaction n inducer molecules bind simultaneously the promoter P:

$$nI + P \underset{k_{-1}}{\overset{k_1}{\rightleftharpoons}} P^*, \tag{3.20a}$$

$$P^* \xrightarrow{k_2} P^* + E. \tag{3.20b}$$

The activation of the promoter by the inducer is considered a reversible reaction and according to the law of mass action in Eq. 3.6 follows the equation:

$$\frac{dP^*}{dt} = k_1 I^n P - k_{-1} P^*. \tag{3.21}$$

Assuming now that the total amount of promoter P_T is constant:

$$P_T = P + P^*, \tag{3.22}$$

Eq. 3.21 becomes:

$$0 = k_1 I^n P_T - (k_1 I^n + k_{-1}) P^*, \tag{3.23}$$

arriving at the following expression:

$$P^* = \frac{k_1 I^n}{k_1 I^n + k_{-1}} P_T = \frac{(\frac{I}{K_H})^n}{1 + (\frac{I}{K_H})^n} P_T. \tag{3.24}$$

Equation 3.24 defines a kinetic law known as *Hill function* where K_H is the Hill constant defined as $K_H^n = k_{-1} / k_1$ and n is the Hill coefficient or *cooperativity index*. The Hill function is generally used to model binding kinetics between a ligand and a receptor including the following cases:

1. *Ligand*: a transcription factor, → *receptor*: a promoter;
2. *Ligand*: an inducer, → *receptor*: a promoter;
3. *Ligand*: an effector, → *receptor*: an enzyme (allosteric regulation);
4. *Ligand*: a drug, → *receptor*: a target protein.

When $n > 1$ the binding is considered cooperative. Empirical values for K_H and n are experimentally determined by tracing the so-called "dose-response" curve.

Previous reactions in Eq. 3.20 considered that enzyme concentration E depended only on the activation of the promoter. However, enzyme concentration is also affected by other effects:

1. **Promoter leakage**, which means that some basal level of enzyme is expressed even in the case of an inactivated promoter;
2. **Enzyme decay**, that accounts for enzyme degradation.

Therefore, the model should include additional reactions reflecting such behavior:

$$nI + P \underset{k_{-1}}{\overset{k_1}{\rightleftharpoons}} P^*, \tag{3.25a}$$

$$P^* \xrightarrow{k_2} P^* + E, \tag{3.25b}$$

$$\varnothing \xrightarrow{k_{leak}} E, \tag{3.25c}$$

$$E \xrightarrow{k_d} \varnothing. \tag{3.25d}$$

According to the model in Eq. 3.25, the enzyme concentration dynamics will obey the following equation:

$$\frac{dE}{dt} = k_{lk} + V \frac{(\frac{I}{K_H})^n}{1 + (\frac{I}{K_H})^n} - k_d E, \tag{3.26}$$

where $V = k_2 P_T$ is the maximal expression rate for the enzyme E.

In the case that the inducer I introduces inactivation or repression of gene expression, the Hill function in Eq. 3.26 changes as follows:

$$\frac{dE}{dt} = k_{lk} + V \frac{1}{1 + (\frac{I}{K_H})^n} - k_d E. \tag{3.27}$$

In addition, we might want to consider transcription explicitly in the model by adding the production of the enzyme E mRNA (messenger RNA) transcript:

$$\frac{dmRNA}{dt} = k_{lk} + V \frac{(\frac{I}{K_H})^n}{1 + (\frac{I}{K_H})^n} - k_d \cdot mRNA, \tag{3.28}$$

$$\frac{dE}{dt} = k_1 \cdot mRNA - k_{-1} E. \tag{3.29}$$

However, in our case we can assume that protein production occurs at a time scale that is faster than the main pathway dynamics of transient concentrations of substrates and products. Typical rates[6] for transcription and translation in bacteria are as follows:
- **Transcription rate:** ≈ 50 bases/s;
- **Translation rate:** ≈ 10 aa/s (amino acids per second).

---

6   Values taken from ▶ http://bionumbers.hms.harvard.edu/

We next define our model for the inducible promoter dynamics in COPASI. Rather than including the several reactions in Eq. 3.25, the rate law will be summarized through the Hill function given in Eq. 3.26. To that end, we will follow the steps:

1. Define two chemical species: enzyme (E) and inducer (I). The initial concentration of the enzyme will be of 0. We will assume that the inducer is added at $t = 0$;
2. Define a new rate law in Functions with the following formula (equivalent to Eq. 3.25):

■■ **Code 3.1 Formula definition in COPASI for the rate law given in Eq. 3.25.**

```
k1 + V*(I/Kh)^n/(1+(I/Kh)^n) -kd*E
```

3. Create a single reaction with defined as I ->E + I with the new rate law given in Code 3.1.

▣ Table 3.3 shows the concentrations of each species. At time $t = 0$ the promoter is fully inactivated, enzyme initial concentration is 0 (although we could add here a basal value due to leakage) and 1 mM of the inducer is added to the media. Parameters of the rate laws are shown in ▣ Table 3.4. As in the previous example for Michaelis-Menten kinetics, we have selected pre-defined rate laws from COPASI. ▣ Figure 3.5 shows the dynamics response of the promoter and the enzyme concentration for a duration of 60 s. As we can see from the plot, the enzyme concentration reaches a steady-state of approximately 1 mM.

▣ **Table 3.3** Initial concentrations for the simulation of the promoter activation

Name	Species	Concentration
PAL	E	0 mM
Inducer	I	10 mM
trans-cinnamate		0 mM

▣ **Table 3.4** Parameters of the rate laws

Definition	Name	Value	Unit
Leakage	k_1	10^{-7}	mmol/(ml*s)
Maximal expression rate	V	10	mmol/(ml*s)
Hill coefficient	K_h	5	mmol/ml
Cooperativity coefficient	n	4	
Degradation	k_d	0.1	s^{-1}

3

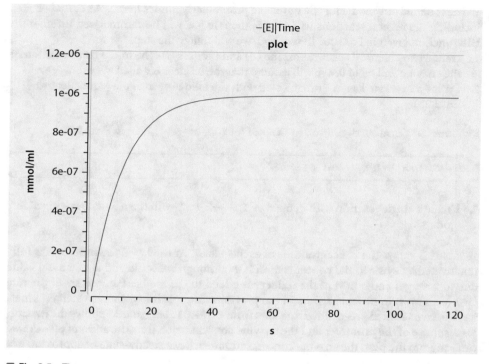

□ **Fig. 3.5**   Time course response of activated promoter concentration (blue curve) and PAL concentration (red curve)

3.5 Modeling Pathway Dynamics

We finally put together the full set of reactions combining the reactions in Eq. 3.9 and 3.25 to define the full model:

$$E + S \; \underset{k_r}{\overset{k_f}{\rightleftharpoons}} \; ES \xrightarrow{k_{cat}} E + P, \tag{3.30}$$

$$nI + P \; \underset{k_{-1}}{\overset{k_1}{\rightleftharpoons}} \; P^*, \tag{3.31}$$

$$P^* \xrightarrow{K_2} P^* + E, \tag{3.32}$$

$$\varnothing \xrightarrow{k_{leak}} E, \tag{3.33}$$

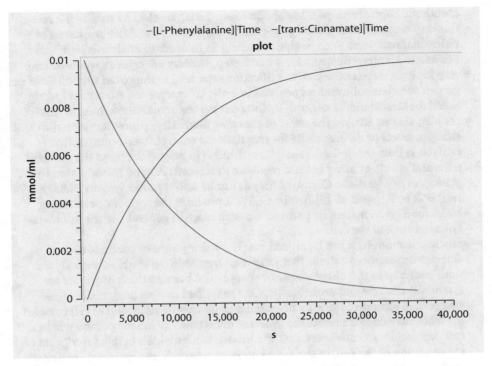

■ **Fig. 3.6** Time course response of the induced expression of PAL enzyme for substrate *L*-phenylalanine concentration (red curve) and product trans-cinnamate concentration (blue curve)

$$E \xrightarrow{k_d} \varnothing. \tag{3.34}$$

■ Figure 3.6 shows the time course responses of the concentrations of substrate and product for a simulation of 10 hours in COPASI. Compared with the dynamic response in ■ Fig. 3.4, where concentration of the enzyme PAL was kept constant at 10 μM, enzyme concentration is controlled here by the activated promoter and decay rate (see ■ Fig. 3.5). The main effect of the regulation of enzyme expression is a decrease in the production rate. If in ■ Fig. 3.4 half of the maximum concentration of the product was achieved after approximately 15 min, reaching same levels of accumulation took approximately 1 h 30 min for the induced response given by the activated promoter in ■ Fig. 3.6. Even if the model was a simplified model and multiple assumptions needed to be introduced, the results of the simulations can already provide some initial hints about the effect of the different parameters on the dynamic response and ways to alleviate bottlenecks.

The described example presented a basic approach for modeling pathway dynamics. Interestingly, this problem can be approached in more detail through several advanced techniques, which will not be covered in detail in this textbook but links to relevant references will be provided here for the interested readers. These techniques include:

— **Metabolic control analysis**: Metabolic control analysis (MCA) studies the relationship between steady-state properties of the pathway and the properties of individual reactions. Such analysis can help us in order to analyze which Michaelis-Menten constants for a multi-step pathway are critical, where are the rate-limiting steps and which modifications can lead to improved titers. Therefore, we can use metabolic control analysis in order to analyze the effect of transcriptional and translational control. By introducing the equations defined in this chapter, we can analyze the effect of choosing alternative promoters with different strength levels or different RBS for translational control. Metabolic control analysis is therefore a useful analytic technique to help understand the different temporal effects of using genetic regulatory elements. A good introduction to the principles of Metabolic Control Analysis can be found in the *Systems Biology* textbook by Klipp et al. [5], whilst COPASI provides the tools for performing MCA for a given metabolic pathway or, even more in general, for the full metabolic network of the cell.

— **Stochastic models**: It can be argued that the assumptions of continuous and deterministic concentrations that were used for modeling the biochemical reactions analyzed in this chapter are not always valid because these processes are essentially discrete and probabilistic. Our simplified models are ignoring many effects such as non-uniformity of spatial distribution of molecules in cell compartments or they do not capture the inherent uncertainty of kinetic parameters in a cell population. In such cases, random models can provide a better description of the dynamics of the cell. In the most common approach, the variables are discrete molecule numbers. A convenient way to simulate those stochastic systems is the Gillespie algorithm [1]. Repeated application of the algorithm allows statistical analysis of the responses. This allows taking into account the uncertainties that are inherent to these processes and to analyze their macroscopic effects. More details about the different types of stochastic simulation and spatial models can be found in Klipp et al. [5].

— **Pathway dynamic regulation**: Transcriptional and translational control through promoters and RBS allow tuning gene expression under some designated conditions. However, cell environmental conditions are variable. This is especially challenging when pathway regulatory elements of an engineered strain are optimized for small-scale lab experiments but are intended to work in the same way in large industrial fermenters. An alternative solution is pathway dynamic regulation, i.e., the introduction of genetic feedback circuits that sense the actual levels of production of the target and regulate the gene expression. Many examples of feedback regulation circuits exist in nature, such as the ones that regulate amino acid production in many organisms. Feedback circuits require of a biosensor system, i.e, some genetic system that is responsive to the controlled signal, generally the concentration of some chemical species [7]. By fusing the biosensor to the promoter circuit, it is possible to regulate the levels of expression of some enzyme depending on the availability of the desired product. Dynamic regulation, however, introduces several engineering challenges such as stability and robustness of the genetic circuit. A review on the subject of dynamic regulation in synthetic biology can be found in a recent review in the engineering journal *IEEE Control Systems* [3].

3.6 Problems

? 3.1 Alternative enzyme kinetics.

✓ Go to BRENDA[7] database and select kinetic parameters from at least three different sources for the PAL enzyme. Simulate them using COPASI in a similar way as shown in ◙ Fig. 3.4. Do you see significant changes in the dynamic response?

? 3.2 Finding enzyme kinetics in databases.

✓ Go to BRENDA database and select kinetic parameters for at least 2 sources for transaminase enzymes (aspartate transaminase EC 2.6.1.1; alanine transaminase EC 2.6.1.2). Simulate them using COPASI in a similar way as shown in ◙ Fig. 3.4. Do you see significant changes in the dynamic response between the EC classes?

? 3.3 Simulate cascade kinetics.

✓ We consider now a two step pathway consisting of the enzymes PAL (as in Problem 3.1) and 4CL (EC 6.2.1.12). Go to BRENDA database and select kinetic parameters for a pair of enzymes from same organism source having kinetics data in the database. Simulate them using COPASI in a similar way as shown in ◙ Fig. 3.4. Which of the two enzymes is the bottleneck of the pathway in terms of reaction rates?

7 ▶ https://brenda-enzymes.org

? 3.4 Promoter library.

✓ ◻ Table 3.4 defined the rate laws for a promoter-activated enzyme expression. Generate a library of 5 promoters by changing the parameters and simulate the responses. *Hint:* use the Hill coefficient parameter in order to modify the library.

? 3.5 Pathway tuning.

✓ As in Problem 3.4, consider promoter rate laws for a cascade pathway consisting of the PAL and 4CL reactions (Problem 3.3). Simulate the dynamic response of the intermediate and target metabolites and assess the effect on the response depending on the rate laws of each promoter. Which promoter allows better tuning of the dynamic response of the pathway?

References

1. Faulon, J.L., Carbonell, P.: Reaction network generation. In: Handbook of Chemoinformatics Algorithms, pp. 317–342. CRC Press, Boca Raton (2010). http://www.crcpress.com/product/isbn/9781420082920
2. Hoops, S., Sahle, S., Gauges, R., Lee, C., Pahle, J., Simus, N., Singhal, M., Xu, L., Mendes, P., Kummer, U.: COPASIa COmplex PAthway SImulator. Bioinformatics **22**(24), 3067–3074 (2006). https://doi.org/10.1093/bioinformatics/btl485
3. Hsiao, V., Swaminathan, A., Murray, R.M.: Control theory for synthetic biology: recent advances in system characterization, control design, and controller implementation for synthetic biology. IEEE Control Syst. **38**(3), 32–62 (2018). https://doi.org/10.1109/MCS.2018.2810459
4. Karr, J.R., Sanghvi, J.C., Macklin, D.N., Gutschow, M.V., Jacobs, J.M., Bolival, B., Assad-Garcia, N., Glass, J.I., Covert, M.W.: A whole-cell computational model predicts phenotype from genotype. Cell **150**(2), 389–401 (2012). https://doi.org/10.1016/j.cell.2012.05.044; http://www.citeulike.org/user/pablocarb/article/10904155; https://linkinghub.elsevier.com/retrieve/pii/S0092867412007763
5. Klipp, E., Liebermeister, W., Wierling, C., Kowald, A., Lehrach, H., Herwig, R.: Systems Biology: A Textbook. Wiley-Blackwell (2009). https://doi.org/10.3797/scipharm
6. Salis, H.M.: The ribosome binding site calculator. In: Methods in Enzymology, vol. 498, pp. 19–42. Elsevier, Amsterdam (2011). https://doi.org/10.1016/B978-0-12-385120-8.00002-4
7. Shi, S., Ang, E.L., Zhao, H.: In vivo biosensors: mechanisms, development, and applications. J. Ind. Microbiol. Biotechnol. **45**(7), 491–516 (2018). https://doi.org/10.1007/s10295-018-2004-x

Further Reading

Useful introductions to **enzyme kinetics** and the Michaelis-Menten model can be found in biochemistry textbooks:
Berg, J.M., Tymoczko, J.L., Stryer, L.: Biochemistry. Freeman (2011)
Nelson, D.L., Cox, M.M.: Lehninger Principles of Biochemistry. Freeman (2013)

The following textbooks provide excellent and comprehensive introductions to the field of **modeling and simulating pathway dynamics**:
Marchisio, M.A.: Introduction in Synthetic Biology: About Modeling, Computation, and Circuit Design. Springer (2018)
Klipp, E., Liebermeister, W., Wierling, C., Kowald, A., Lehrach, H., Herwig, R.: Systems Biology: A Textbook. Wiley-Blackwell (2009)
Wall, M.E.: Quantitative Biology: From Molecular to Cellular Systems. CRC Press (2013)
Munsky, B., Hlavacek, W.S., Tsimring, L.S.: Quantitative Biology: Theory, Computational Methods, and Models. MIT Press (2018)

Modeling Chemical Diversity

4

What You Will Learn in This Chapter

In this chapter, you will learn about ways to model the chemical diversity found in metabolic pathways in nature. Organisms have evolved enzymes, i.e., specialized proteins to carry out chemical transformations that produce the compounds required for life. We have nowadays a good understanding about the mechanisms of natural evolution that allowed the creation of new enzymes and new activities. We are going to model and simulate such behavior by encoding reactions in the same way as we encode a language using words. This will allow us to understand the grammar behind the generation of new reactions. Even more interestingly, we will see how the grammar can be potentially used to enumerate any possible reaction and any possible compound that can be produced in nature. At the end of this chapter, we should have gained a good understanding of the biochemical space that exists in nature.

4.1 Understanding the Mechanisms of Enzyme Innovation

Imagine how was life several millions of years ago. Go as far back in time as to the Precambrian era. Life was then less complex than the way we know it today. Cells carried out basic activities with a rather limited set of proteins. Multiple activities occurred simultaneously through the same proteins, which played a generalist role before becoming more specialized into specific pathways and metabolic functions. Ancient enzymes were generalist, they had the ability of performing multiple functions, i.e., they were highly **promiscuous**. As organisms evolved through the course of time and species evolved on Earth, proteins became more and more specialized [5]. Nowadays, enzymes still show a high degree of promiscuity but perhaps in a least degree than their ancestors. Having such degree of plasticity, such flexibility, is essential for **adaptation**. If for some reason environmental conditions changed, say some toxic compound appeared in the environment, those cells with the ability of evolving an enzyme that can degrade the toxic compound would be the ones that will survive. Enzyme promiscuity therefore provides an evolutionary advantage by enabling enzyme innovation and adaptation.

Enzyme promiscuity and evolvability are keystones in biotechnology [6]. In biotechnology labs, we can accelerate evolution by selectively mutating enzymes so that they can change their original function and start working in a slightly different manner. We will learn more about directed evolution later in ▶ Sect. 9.2. But first, let us have here a look at how to model promiscuous reactions so that we have the tools that will help us in order to model enzyme evolution.

4.2 Knowledge-Based Encodings for Chemical Reactions

Enzymes catalyze chemical transformations. These proteins are nature's basic tools that allow creating chemical diversity by facilitating the chemical transformation of substrates into products. Synthetic biology aims at introducing some modifications to modify the way these reactions work. We can think of at least three basic goals that synthetic biology tries to achieve:

- **Heterologous expression:** Importing the genes of enzymes that are natively found in one organism, i.e., endogenous genes, into another host organism in order to catalyze the original reaction in the host's cell environment. This can be useful in many contexts, especially when the original reaction occurs natively in a host like a plant that is difficult to harvest and therefore producing the target chemical in an industrial organism is advantageous;

— **Enzyme efficiency**: Modifying the original enzyme sequence of amino acids by introducing some mutations that make the reaction more efficient or with better performance;

— **Substrate specificity**: Modifying substrate specificity, taking advantage of enzyme promiscuity in order to look for an enzyme with some increased affinity towards some substrate or to catalyze some reaction.

These goals and many other related ones are the subject of study of **biocatalysis**, a vast discipline in biochemistry that is renowned for its many industrial success stories both *in vitro* and *in vivo*. Going in depth into the subject of biocatalysis is out of the scope of this textbook. However, we would like to provide here a glimpse into some of the tools that are available in order to address biocatalysis problems from the viewpoint of computational pathway design.

The first think that we need is to represent biochemical reactions in a meaningful way that helps for our purposes. Encoding chemical structures in a form that can be entered as a text string for computer processing has a long tradition starting in the 1960s. We cannot review here such long history. We will focus on one the best-known and most popular formats, the SMILES notation (Simplified Molecular Input Line Entry System). SMILES provides intuitive and relatively simple input rules. The full description can be downloaded from the web site of Daylight Chemical Information Systems Inc.[1] A summary of the basic rules is shown in ▶ Box 4.1.

Box 4.1
Summary of basic SMILES rules:
— Start with an atom of the chemical structure and write down its symbol within bracket squares (brackets can be omitted for most commonly found atoms in organic structures).
— Write down a symbol to represent the order of the bond with a neighboring atom: -, =, #, : for single, double, triple or aromatic bond respectively. Single bonds can be generally omitted.
— Parenthesis represent branching points (more than one neighboring atom).
— Ring structures are identified by breaking the bond in the ring and labeling the two atoms with a unique digit number.
— Hydrogen atoms can be included explicitly or omitted.
— Anticlockwise stereochemistry for atoms is represented by "@", clockwise by "@@".

An extension of SMILES is the SMARTS [4] notation, which has been developed to represent generalized chemical patterns and substructures. We will make use of SMARTS in order to write down generic chemical rules.[2] Reactions can be represented using SMILES as well. Each reactant set is represented as SMILES strings separated by " . " and the two

---

1 ▶ http://www.daylight.com/
2 I invite the reader to learn in detail the rules about SMARTS to understand the big possibilities of using this common representation for chemical transformations.

strings are joined together with the symbol ">>". A further extension, called SMIRKS allows representing generic reactions using the SMARTS notation. In order to avoid ambiguity, each atom in the substrate set might be mapped to its corresponding atom in the product using atom numbers. Atom numbers are represented by a number following a colon within the square brackets.

The full rules about SMILES and SMARTS are out of the scope of this textbook. The most important idea to keep in mind is that virtually any known chemical transformation can be represented through SMIRKS, and therefore can be expressed in a well-controlled language that can be processed by the software or, using computer terms, *can be compiled*. We first practice in order to be able to write down some molecules in SMILES notation that will be combined into a SMIRKS reaction.

▪ Table 4.1 shows some examples of chemical structures and their corresponding SMILES notation. Besides the rules in ▶ Box 4.2, we see that a carbon atom in lower case "c" represents an aromatic carbon and that ionization can be explicitly given by using the sign symbols "+", "−".

▪ **Table 4.1**   Examples of chemical structures written in SMILES notation

Diagram	SMILES	Formula	Name
H+	[H+]	H	Proton
CH_4	C	CH_4	Methane
H_2O	O	H_2O	Water
OH	CCO	C_2H_6O	Ethanol
	C1CCCCC1	C_2H_{12}	Cyclohexane
	c1ccccc1	C_6H_6	Benzene
	c1ccccc1[N+](=O)[O-]	$C_2H_5NO_2O$	Nitrobenzene
OH NH_2	N[C@@H](C)C(=O)O	$C_3H_7NO_2$	L-alanine
HO NH_2 OH	N[C@@H](Cc1ccc(O)cc1)C(O)=O	$C_9H_{11}NO_3$	L-tyrosine

The user is not expected to input the chemical information in a unique way, i.e., there exists multiple SMILES representations of a molecular structure depending for instance on the starting atom. However, computers should be able to convert the SMILES string into a unique representation so that there is a one-to-one relationship between the structure and its internal representation. In order to avoid ambiguity, canonical SMILES and InChI Code (see ▶ Box 4.2) are two string notations of chemical structures that are generated through a non-ambiguous set of rules. Again, we are not expected to write down such codes by hand, but rather to generate them through the computer and use them to perform chemical comparisons, databases searches, etc.

Box 4.2
The **IUPAC International Chemical Identifier (InChI)** is a standard to encode molecular information of chemical substances. Every InChI identifier starts with the string "InChI=" followed by the version and the letter "S" for standard. The chemical information is provided in up to six layers that are separated by the delimiter "/". The main layers and sublayers are:

1. **Main layer:**
    a. **Chemical formula,** which is the only layer that must occur in every InChI.
    b. **Atom connections** describes how atoms are connected by bonds, prefix "c".
    c. **Hydrogen** described how hydrogens are connected to other atoms, prefix "h".
2. **Charge:** either "p" for protons or "q" for charge.
3. **Stereochemistry:** "b" for double bonds, "t" for tetrahedral, etc.

For instance, ethanol (SMILES: CCO) is encoded as `InChI=1S/C2H6O/c1-2-3/h3H,2H2,1H3`. The **InChIKey** is a 27-character hashed version (hashing will be discussed later in ▶ Sect. 5.2) of the full InChI that allows easy database and web search of chemical compounds.

Now that we have learned the basics of chemical notation using SMILES, our next step is to actually enter the chemical information into the computer so that it can be processed. For that purpose, as in the rest of this textbook we are going to use libraries in Python. The most popular library for computational processing of chemical structures, a scientific field generally known as Chemoinformatics, is `RDKit` (see ▶ Box 4.3). One basic operation that we can do with this library is to create a `mol` object, i.e., an internal representation of a molecule in `RDKit` that can be then used for further processing using a large collection of tools. In the following example, a `mol` object is created for methane (CH_4) using the `MolFromSmiles` routine:

Box 4.3
`RDKit` is an open-source library of Chemoinformatics tools in Python that has been developed over the years and has an important community of users and developers. The number of tools and calculations that can be done using `RDKit` is quite extensive and growing everyday. Some of them:

- Input/output chemicals in multiple formats.
- Substructure and similarity search.
- Chemical transformations and reactions.
- Computing chemical properties and descriptors.

- 2D depiction.
- Fingerprint calculations for screening and machine learning.
- Integration with other tools such as KNIME [3] (workflows), Django (web server) and PostrgreSQL (database).

The module IPythonConsole allows drawing molecules directly in the Jupyter notebook. The reader interested in learning more about using RDKit in Chemoinformatics should follow the tutorials available in the RDKit website (▶ http://rdkit.org).

RDKit provides multiple modules for computing chemical properties and descriptors. For instance, in order to compute the molecular weight we use the function MolWt from the module Chem.Descriptors:

■■ Code 4.1 Define a mol object in RDKit.

```
from rdkit import Chem
from rdkit.Chem.Draw import IPythonConsole
methan = 'C'
mol = Chem. MolfromSmiles (methane)
display (mol)
```

■■ Code 4.2 Compute the molecular weight using RDKit.

```
from rdkit import Chem.Descriptors
print(Descriptors. MolWt (mol))
```

Next we are going to create a mol object containing a biochemical reaction. We can start with L-tyrosine ammonia lyase (TAL), EC 4.3.1.25, a reaction that converts tyrosine into p-coumaric acid (see ◻ Fig. 4.1), an important pathway precursor of many valuable natural products.

As mentioned previously, reactions are simply defined using SMILES by concatenating the reactants through the " . " symbols at each side of the reaction, separated by the symbol ">>". Code 4.3 defines the SMILES for the substrates and the products of TAL and then concatenates them in order to define the reaction.

◻ **Fig. 4.1** L-tyrosine ammonia lyase (TAL) reaction. Rendering of TAL is performed by RDkit

■■ Code 4.3 Define a reaction using SMILES.

```
tyrosine = 'N[C](Cc1ccc(O)cc1)C(O)=O'
ammonia = 'N'
coumarate = 'OC(=O)C=Cc1ccc(O)cc1'
tal = tyrosine+'>>'+coumarate+'.'+ammonia
r = Chem.rdChemReactions.ReactionFromSmarts(tal)
display(r)
```

By simply inspection of the SMILES strings or by looking at a reaction depiction, we can easily identify common chemical groups that remained unchanged through the transformation from L-tyrosine to p-coumaric acid as well as others that were modified through the transformation. In that way, the previous reaction can be converted into a **reaction rule** where the unmodified chemical part of the reaction is substituted by a generic group. Using SMARTS notation, the most generic representation would be a wildcard "*" meaning any possible atom (see Code 4.4). The resulting rule, shown below in ◘ Fig. 4.2, is a generic representation of the reaction where the phenolic group "Cc1ccc(O)cc1" has been replaced by a generic group.

■■ Code 4.4 Define a reaction using SMARTS.

```
R2 = '*'
tyrosine = 'N[C]({})C(O)=O'.format(R2)
ammonia = 'N'
coumarate = 'OC(=O)C={}'.format(R2)
tal = tyrosine+'>> '+coumarate+'.'+ammonia
r = Chem.rdChemReactions.ReactionFromSmarts(tal)
display(r)
```

A chemoinformatics tool such as RDKit allows us to work with reaction rules and to test if a reactant can be transformed by the reaction rule. However, in order to do that properly RDKit requires the **atom-atom mapping** of the reaction. In other words, the program needs to know where each atom of the left reactants is mapped on the right reactants through the reaction. This allows limiting the combinatorial complexity that otherwise would be present in the calculation of the reaction rule. In simple cases, like in the

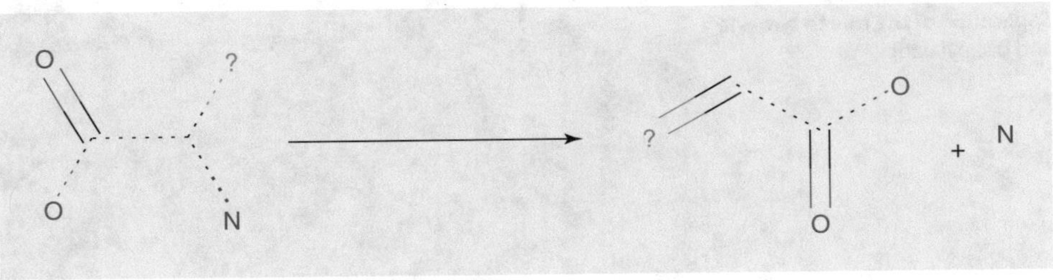

◘ Fig. 4.2 Reaction rule derived from TAL

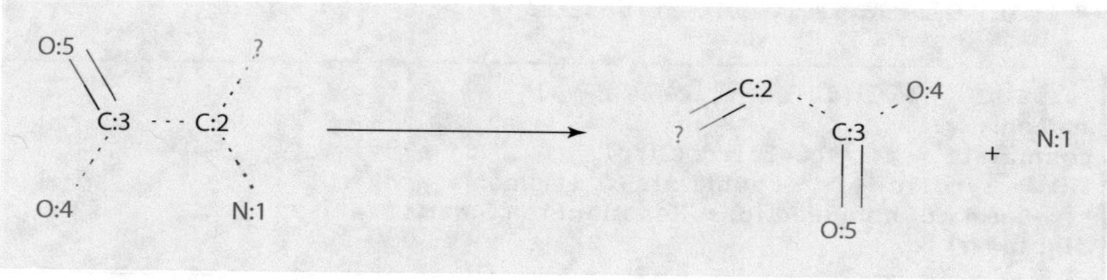

◻ Fig. 4.3 Atom-atom mapping of the TAL rule

example below for TAL in Code 4.5 and shown in ◻ Fig. 4.3, we can perform manually the atom-atom mapping, in more complex cases, we will need to make use of an automated tool that would perform the mapping like the Reaction Decoder Tool (RDT) developed at the European Bioinformatics Institute (EBI) [8].

■■ Code 4.5 Define a generic reaction in SMARTS.

```
R2 = '[*:6]'
tyrosine = '[N:1][C:2]({})[C:3]([O:4])=[O:5]'.format(R2)
ammonia = '[N:1]'
coumarate = '[O:4][C:3](=[O:5])[C:2]={}'.format(R2)
tal = tyrosine+'>>'+coumarate+'.'+ammonia
r = Chem.rdChemReactions.ReactionFromSmarts(tal)
display(r)
```

Once we have the reaction rule defined using the atom-atom mapping, we can test if the rule can be triggered by multiple substrates. RDKit uses the command RunReactants to calculate all possible reactants that can be produced by the given substrates as long as their chemical structures fit into the rule (see Code 4.6). One alternative substrate to test for the rule is phenylalanine, an aromatic amino acid that differs in tyrosine only on the hydroxyl group in the ortho position of the aromatic ring. Since the rule focuses around the reaction center, far from the aromatic ring, we might expect that the rule will accept both substrates. As shown in ◻ Fig. 4.4, the rule generated a product both for L-tyrosine (p-coumaric acid) and for L-phenylalanine (cinnamic acid).

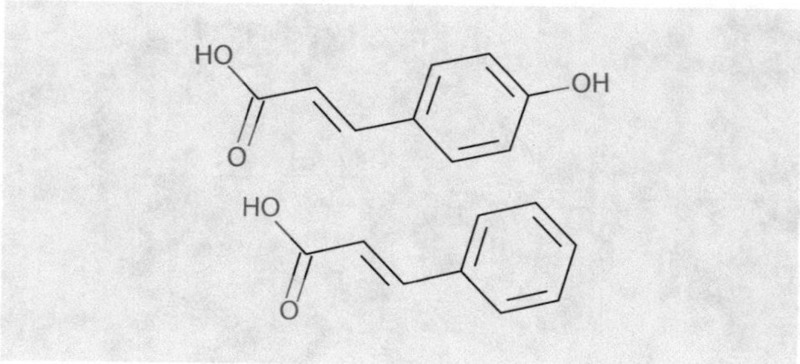

◻ Fig. 4.4 Products of the application of the rule to tyrosine and phenylalanine

■■ **Code 4.6 Reaction enumeration using RDKit.**

```
tyrosine  =  'N[C](Cc1ccc(O)cc1)C(O)=O'
phenylalanine  =  'NC(Cc1ccccc1)C(O)=O'
mtyrosine  =  Chem.MolFromSmiles(tyrosine)
mphenylalanine  =  Chem.MolFromSmiles(phenylalanine)
x  =  r.RunReactants((mtyrosine,))
display(x[0][0])
x  =  r.RunReactants((mphenylalanine,))
display(x[0][0])
```

4.3 Modeling Enzyme Promiscuity

In the previous section we saw an example of modeling a property that is often observed in natural enzymes, i.e., **enzyme promiscuity**. Enzyme promiscuity is the ability of enzymes to accept multiple substrates or catalyze multiple reactions. Such ability is recognized as a major drive allowing innovation in nature. Enzyme promiscuity can be also used within a biotechnological context in order to discover new capabilities in enzymes. An important discovery was that employing the type of generic representation of reactions or "rules" discussed in previous ▶ Sect. 4.2 allows reproducing a similar behavior as the one that is observed in enzymes [1]. Generally, enzymes will display a higher or lower level of specificity to the substrate. This is something that we can represent by **lowering or increasing the number of atoms considered around the reaction center**.

For instance, in the following example, we will write the biochemical transformation for TAL starting from the atoms in the reaction center and then moving away extending the representation up to the full molecule. In all cases, we check that the pair of substrates in our example, *L*-tyrosine and *L*-phenylalanine can be processed by the reaction. When we extend the reaction representation to the full set of atoms in tyrosine including the hydroxyl group in the aromatic ring, the reaction would not be triggered for phenylalanine because of the difference in such chemical group. In ▶ Box 4.4 there is a summary description of some of the considerations that are needed in order to write down reaction rules using SMARTS (SMIRKS) representation.

Box 4.4

Summary of steps in order to generate reaction rules:

- Atoms need to mapped between reactants, at least the ones that can accept multiple substitutions. Software such as RDT [8], EC-BLAST [7] or MetaCyc [2] can help us to do the atom-atom mapping.
- Identify the reaction center, i.e., the atoms and bonds that are transformed through the conversion. Again, the previous software can help us in finding the reaction center.
- Initially, write down the most basic rule corresponding to the basic transformation around the reaction center. Add constraints following the SMARTS definitions to the type of bonds and atoms that are allowed.
- For simplicity, hydrogens can be left implicit, although more complex cases might require the full description of hydrogens (protonation and ionization states) for each atom.
- Be careful with aromatic rings. Molecules are often represented in Kekulé form using the Kekulization algorithm. However such algorithm might bring to different results when looking at atom environments around the reaction center. Therefore, it is recommended to explicitly describe aromatic atoms as in the example.

Initially, Code 4.7 focuses on only the atoms and bonds that are transformed through the biochemical transformation (◼ Fig. 4.5).

■■ Code 4.7 Define the TAL reaction center.

```
x_tyrosine = '[N:1][C:2]([C:3])'
x_ammonia = '[N:1]'
x_coumarate = '[C:2](=[C:3])'
tal = x_tyrosine+'>>'+x_coumarate+'.'+x_ammonia
r0 = Chem.rdChemReactions.ReactionFromSmarts(tal)
display(r0)
```

The rule is then validated in Code 4.8 by checking that coumarate can be produced from tyrosine. Moreover, we will use the method `GetSubstructMatch` in order to highlight the reaction center in both reactants (see ◼ Fig. 4.6).

■■ Code 4.8 Run multiple reactants through the generic reaction.

```
def displayMainReactants(r):
    s_tyrosine = 'N[C](Cc1ccc(O)cc1)C(O)=O'
    tyrosine = Chem.MolFromSmiles(s_tyrosine)
    tyrosine.GetSubstructMatch(Chem.MolFromSmarts(x_tyrosine))
    x = r.RunReactants((tyrosine,))
    coumarate = x[0][0]
    coumarate.GetSubstructMatch(Chem.MolFromSmarts(x_coumarate))
    display(tyrosine)
    display(coumarate)
displayMainReactants(r0)
```

Similarly, in Codes 4.9, 4.10 and 4.11 the rule can be extended to a wider atomic environment around the reaction center, as shown in ◼ Figs. 4.7 and 4.8 (rule 1), 4.9 and 4.10 (rule 2), and 4.11 and 4.12 (rule 3), respectively.

◼ **Fig. 4.5**   Rule 0

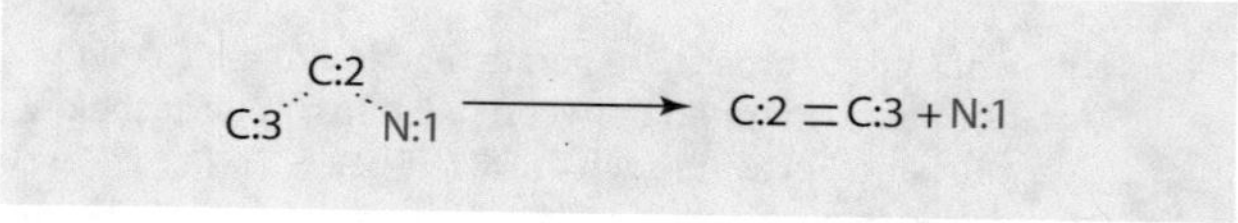

◼ **Fig. 4.6**   Reactants for Rule 0

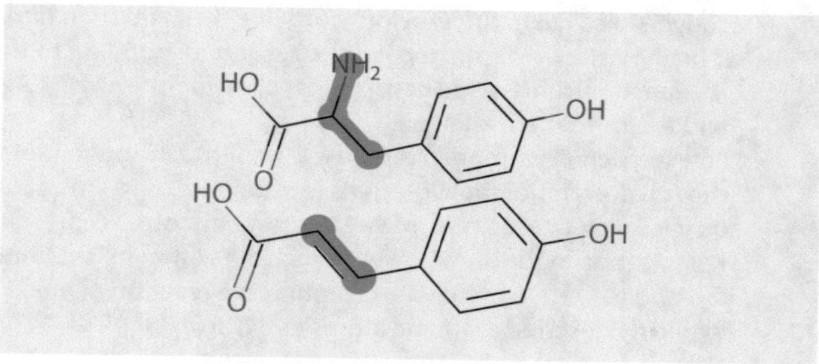

Fig. 4.7 Rule 1

Fig. 4.8 Reactants for rule 1

Fig. 4.9 Rule 2

Fig. 4.10 Reactants for rule 2

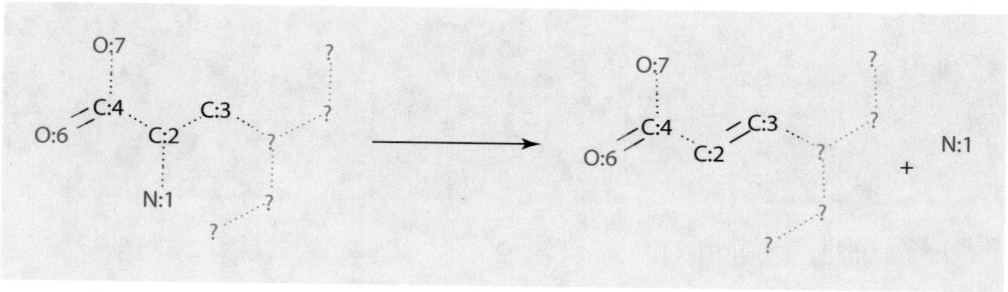

4

Fig. 4.11   Rule 3

Fig. 4.12   Reactants for Rule 3

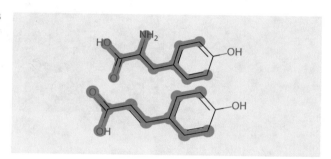

Code 4.9 Using a wider atomic reaction center in for reactant enumeration.

```
x_tyrosine  = '[N:1][C:2]([C:3][*:4])[C:5]'
x_ammonia  = '[N:1]'
x_coumarate  = '[C:2](=[C:3][*:4])[C:5]'
tal = x_tyrosine+'>> '+x_coumarate+'.'+x_ammonia
r1 = Chem.rdChemReactions.ReactionFromSmarts(tal)
display(r1)
displayMainReactants(r1)
```

Code 4.10 Increasing the reaction center diameter.

```
x_tyrosine  = '[N:1][C:2]([C:3][*:5](~[*:8])~[*:9])[C:4](=[O:6])[O:7]'
x_ammonia  = '[N:1]'
x_coumarate  = '[C:2](=[C:3][*:5](~[*:8])~[*:9])[C:4](=[O:6])[O:7]'
tal = x_tyrosine+'>>'+x_coumarate+'.'+x_ammonia
r2 = Chem.rdChemReactions.ReactionFromSmarts(tal)
display(r2)
displayMainReactants(r2)
```

■■ **Code 4.11 Wider reaction center.**

```
x_tyrosine = '[N:1][C:2]([C:3][*:5](~[*:8]~[*:10])~[*:9]~[*:11])
    [C:4](=[O:6])[O:7]'
x_ammonia = '[N:1] '
x_coumarate = '[C:2](=[C:3][*:5](~[*:8]~[*:10]) ~[*:9]~[*:11])
    [C:4](=[O:6])[O:7] '
tal = x_tyrosine+'≫'+x_coumarate+'.'+x_ammonia
r3 = Chem.rdChemReactions.ReactionFromSmarts(tal)
display(r3)
displayMainReactants(r3)
```

4.4 Enumerating Chemical Diversity

As we have seen in previous sections, modeling reaction promiscuity allowed us making hypothesis about pairs of subtrates-products that could potentially be processed by some putative enzyme. Such approach can be extended into enumerating chemical diversity. The objective will be screening a library of chemicals in order to discover how many of them could be processed by the reaction rule.

For instance, in the next step shown in Code 4.12 we will download a library of metabolites from the database HMDB,[3] which is a database specialized in small molecule metabolites found in the human body. The library will be downloaded from its website (in `zip` format) and extracted into a file called `structures.sdf`. SDF is a chemical file format often used in order to store a list of molecular structures and associated properties. SDF files can be read by `RDKit` by using the `SDMolSupplier` function. At the time of this writing HMDB contained 113,983 molecules, which might require some time in order to convert all of them into `mol` objects.

■■ **Code 4.12 Downloading molecular structures from HMDB.**

```
import requests
import zipfile
sdf = 'structures.sdf'
f = 'structures.zip'
url = 'http://www.hmdb.ca/system/downloads/current/'+f
r = requests.get(url, allow_redirects=True)
open(f,'wb').write(r.content)
zip_ref = zipfile.ZipFile(f,'r')
zip_ref.extract(sdf)
zip_ref.close()
cl = Chem.SDMolSupplier('structures.sdf')
clist = [m for m in cl]
```

We will loop now in Code 4.13 through the molecules in order to test how many of them can be converted through the reaction rule into products. Note that this procedure of **reaction enumeration** is going to generate lots of putative novel molecules, which will

4

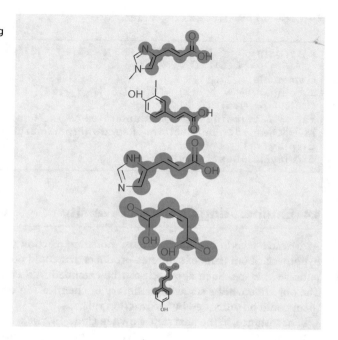

□ **Fig. 4.13** Five examples of enumerated products by running rule 2 in the chemical library

need to be verified experimentally using for instance mass spectrometry or nuclear magnetic resonance (NMR). What we are going to do here is to assume the hypothesis that an enzyme because of its promiscuity is potentially able to catalyze multiple substrates with some level of activity as far as they verify its associated reaction rule. In the following example we will run the library of human metabolites through the rule 2 of the previous section (see Code 4.10 and □ Figs. 4.9 and 4.10) in order to identify how many of them are processed. □ Fig. 4.13 shows the five initially enumerated products in the library.

■■ **Code 4.13 Screen for metabolites that can processed by the generic reaction.**

```
hmr = []
for y in clist:
    try:
        pr = r2.RunReactants( (y,) )
        if len(pr) > 0:
            m = pr[0][0]
            m.GetSubstructMatch(Chem.MolFromSmarts(x_coumarate))
            hmr.append(m)
        if len(hmr) > 10:
            break
    except:
        continue
for l in hmr:
    try:
        Chem.SanitizeMol(l)
        display(l)
    except:
        continue
```

Take Home Message

Enzymes are proteins specialized in catalyzing biochemical transformations in order to drive metabolic processes and to produce all needed metabolites in the cell.

- Enzymes have the ability of accepting multiple substrates or catalyzing multiple reactions, a property that is called enzyme promiscuity and is essential for adaptability through evolution.
- The mechanism of promiscuity can be investigated by looking at common patterns in chemical transformations.
- Using computer representations of reactions, we can analyze common patterns in reactions and infer reaction rules to represent and predict the mechanisms of promiscuity.

4.5 Problems

? **4.1** Reading molecules in `RDKit` in SMILES format.

✓ Create in Python a list or array containing the SMILES strings in ▣ Table 4.1. Loop through them in order to create an array containing a `mol` object per molecule.

? **4.2** Calculating the molecular weight using `RDKit`.

✓ Calculate the molecular weight of the molecules in ▣ Table 4.1 and sort them in ascending order depending on their mass.

? **4.3** Calculating physico-chemical properties using `RDKit`.

✓ Using the `Chem.Crippen` module in `RDKit`, calculate the predicted octanol-water partition coefficient $\log P$, a measure of lipophilicity, of the molecules in ▣ Table 4.1.

? **4.4** Running other amino acids through a reaction rule.

✓ L-DOPA is an amino acid found in the metabolism of many organisms. Test if L-DOPA can be accepted by the reaction rule in Code 4.6. What is the product?

? **4.5** Simulating enzyme promiscuity.

✓ Take the reaction rules for enzyme PAL of increasing specificity in ▣ Figs. 4.6, 4.8, 4.10, and 4.12 and run reactants *L*-tyrosine, *L*-phenylalanine and L-DOPA through them. Are they viable substrates for the enzyme at all the simulated levels?

References

1. Carbonell, P., Planson, A.G., Fichera, D., Faulon, J.L.: A retrosynthetic biology approach to metabolic pathway design for therapeutic production. BMC Syst. Biol. **5**(1), 122 (2011). https://doi.org/10.1186/1752-0509-5-122
2. Caspi, R., Altman, T., Dreher, K., Fulcher, C.A., Subhraveti, P., Keseler, I.M., Kothari, A., Krummenacker, M., Latendresse, M., Mueller, L.A., Ong, Q., Paley, S., Pujar, A., Shearer, A.G., Travers, M., Weerasinghe, D., Zhang, P., Karp, P.D.: The MetaCyc database of metabolic pathways and enzymes and the BioCyc collection of pathway/genome databases. Nucleic Acids Res. **40**(D1), D742–D753 (2012). https://doi.org/10.1093/nar/gkr1014
3. Fillbrunn, A., Dietz, C., Pfeuffer, J., Rahn, R., Landrum, G.A., Berthold, M.R.: KNIME for reproducible cross-domain analysis of life science data. J. Biotechnol. (2017). https://doi.org/10.1016/j.jbiotec.2017.07.028
4. Judson, P.: Knowledge-Based Expert Systems in Chemistry. Theoretical and Computational Chemistry Series. Royal Society of Chemistry, Cambridge (2009). https://doi.org/10.1039/9781847559807
5. Khersonsky, O., Tawfik, D.S.: Enzyme promiscuity: a mechanistic and evolutionary perspective. Annu. Rev. Biochem. **79**(1), 471–505 (2010). https://doi.org/10.1146/annurev-biochem-030409-143718
6. Notebaart, R.A., Kintses, B., Feist, A.M., Papp, B.: Underground metabolism: network-level perspective and biotechnological potential. Curr. Opin. Biotechnol. **49**, 108–114 (2017)
7. Rahman, S.A., Cuesta, S.M., Furnham, N., Holliday, G.L., Thornton, J.M.: EC-BLAST: a tool to automatically search and compare enzyme reactions. Nat. Methods **11**(2), 171–174 (2014). https://doi.org/10.1038/nmeth.2803
8. Rahman, S.A., Torrance, G., Baldacci, L., Martínez Cuesta, S., Fenninger, F., Gopal, N., Choudhary, S., May, J.W., Holliday, G.L., Steinbeck, C., Thornton, J.M.: Reaction Decoder Tool (RDT): extracting features from chemical reactions. Bioinformatics **32**(13), 2065–2066 (2016). https://doi.org/10.1093/bioinformatics/btw096
9. Willighagen, E.L., Mayfield, J.W., Alvarsson, J., Berg, A., Carlsson, L., Jeliazkova, N., Kuhn, S., Pluskal, T., Rojas-Chertó, M., Spjuth, O., Torrance, G., Evelo, C.T., Guha, R., Steinbeck, C.: The Chemistry Development Kit (CDK) v2.0: atom typing, depiction, molecular formulas, and substructure searching. J. Cheminformatics **9**(1), 33 (2017). https://doi.org/10.1186/s13321-017-0220-4

Further Reading

A good introduction to **biocatalysis**:

Grunwald, P.: Biocatalysis. Biochemical Fundamentals and Applications. Imperial College Press (2009)

An interesting discussion on **enzyme promiscuity and evolution**:

Khersonsky, O., Tawfik, D.S.: Enzyme promiscuity: a mechanistic and evolutionary perspective. Ann. Rev. Biochem. **79**(1), 471–505 (2010)

Useful introductions to **chemoinformatics** and associated algorithms can be found in:

Judson, P.: Knowledge-Based Expert Systems in Chemistry. Theoretical and Computational Chemistry Series. Royal Society of Chemistry, Cambridge (2009)

Gasteiger, J., Engel, T. (eds.): Chemoinformatics. Wiley-VCH Verlag GmbH & Co. KGaA, Weinheim, FRG (2003)

Faulon, J.L., Bender, A.: Handbook of Chemoinformatics Algorithms. Chapman & Hall/CRC (2010)

More details about the implementation of chemoinformatics algorithms are available at the sites for the software packages:

The RDKit Python library: http://rdkit.org

The CDK [9] Java library: https://cdk.github.io/

An insightful discussion about **chemical space enumeration**:

Reymond, J.L., Ruddigkeit, L., Blum, L., van Deursen, R.: The enumeration of chemical space. Wiley Interdiscip. Rev.: Comput. Mol. Sci. **2**(5), 717–733 (2012)

Metabolic Pathway Discovery

Contents

Enzyme Discovery and Selection

© Springer Nature Switzerland AG 2019
P. Carbonell, *Metabolic Pathway Design*, Learning Materials in Biosciences,
https://doi.org/10.1007/978-3-030-29865-4_5

5

What You Will Learn in This Chapter

Enzymes have the ability of catalyzing reactions. Each organism has evolved its very own version of an enzyme. Having to select an enzyme sequence for a target reaction is not always straightforward. In some cases, we may have hundreds of variants coming from different organisms to choose from and with little information about the best choice. In other cases, no known enzyme sequence catalyzing the desired reaction would be known. In this chapter, you will learn some basic techniques based on homology modeling both of the sequence and the chemical reaction allowing to guide us when selecting enzyme sequence candidates. Based on such approaches, at the end of this chapter we will discuss how to expand natural capabilities of enzymes through promiscuity.

5.1 Enzyme Discovery Through Sequence Homology

The minimal information of an enzyme is its sequence. A well-known fact is that enzyme sequences with high degree of similarity will most likely be functionally similar. A change of sequence should have an impact on the observed enzymatic activity, on its thermostability, folding, etc. Directed evolution allows experimentally exploring the effects of random mutations in enzyme function (see ▶ Box 5.1). In metabolic pathway design, when confronted with some sequence of unknown function or in the case that we want to find functional alternatives to some parent enzyme sequence our first attempt would be to **search for similar sequences** in databases. We can BLAST[1] our sequence against the full set of sequences in UniProt[2] or in NCBI,[3] or we can focus in some organism or taxa in order to find homologues.

Box 5.1   Screening and Selection Through Directed Evolution

Rather than selecting the best sequence from the catalog of enzymes that are available in natural organisms, a different approach consists on evolving enzymes with the desired activity starting from a parent sequence. Unlike the rational design approach seen in ▶ Sect. 5.1, directed evolution allows us to perform the enzyme selection in the laboratory by mimicking the procedure found in evolution. Directed evolution relies less on mechanistic and structural information, but on high throughput screening and selection from a pool of protein variants. It consists of iterative rounds of:

- **Mutagenesis:** creating a library of variants of the target gene;
- **Selection:** isolating variants with the desired function;
- **Amplification:** generating a template for the next round.

This approach has been successfully used to engineer enzymes [2], opening up new possibilities for producing valuable chemicals through synthetic biology [15].

For instance, consider again the biochemical step catalyzed by the Phenylalanine ammonia lyase (PAL) shown in ▪ Fig. 3.2. In order to discover enzyme sequences with

---

1   BLAST is a widely-used algorithm for comparing sequences.
2   ▶ https://www.uniprot.org/blast/
3   ▶ https://www.ncbi.nlm.nih.gov/BLAST/

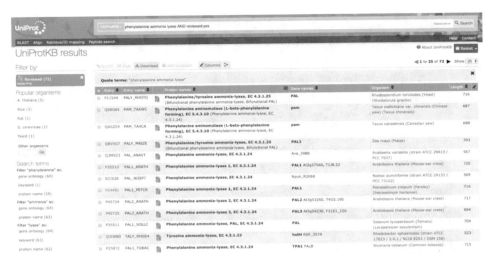

◼ Fig. 5.1   A selection of phenylalanine ammonia lyase curated sequences in UniProt

such activity nothing as easy as performing a search in UniProt for "Phenylalanine ammonia lyase". At the time of this writing, the output of such query search produced 72 reviewed results, i.e., sequences that have been manually revised for their associated experimental evidence. The output for the first sequences in the results of the query is shown in ◼ Fig. 5.1.

UniProt allows us to perform a **multiple sequence alignment** (MSA) of the resulting sequences, which will inform about which residues or regions are typically more conserved. **Conserved regions** are generally assumed to be associated with some function of the protein. This does not mean necessarily that the functional role of the residues is directly linked to the catalytic activity of the enzyme. They can play for instance a structural role in order to serve as a backbone scaffold. Therefore, rather than focusing on the full enzyme sequence, a more meaningful search would be to look for those sequence **motifs and domains** that are common to some specific enzyme function. There exist several classification systems of protein sequences (see ▶ Box 5.2), most of them generally linked from the individual entries of the Uniprot page of the protein sequence. For instance, one popular approach is to use PFam, which is a database of domains that are common to different protein functions. These classification systems will give us different perspectives about the enzyme sequences and often will provide the possibility of classifying a sequence of unknown function. We will see in ▶ Chap. 8 how the different features of enzyme sequences can be ranked in order to prioritize designs in an engineered metabolic pathway.

Another major approach for sequence selection is by exploring **protein structure**. The Protein Data Bank (PDB)[4] contains more than 100,000 protein structures that have been determined by X-ray crystallography, nuclear magnetic resonance (NMR) or electron miscroscopy (EM). Such rich resource of information is helpful for enzyme sequence selection, especially when the structure of multiple variants or homologues of the same enzyme family are available. For those sequences of interest in a given enzyme family for which structure is not available, we can rely on software that predicts the structure based

5

> **Box 5.2   How Proteins Are Classified?**
>
> There exists many ways of classifying protein sequences into classes or families. Each one uses a different criteria or approach generally associated with some automated method for classification but also involving community-based manual curation:
>
> — The simplest approach is **sequence similarity** in order to cluster the sequences into groups. This approach is the one used in order to define *non-redundant* sets of sequences like the one in **UniRef** [18], which uses the highly efficient CD-HIT algorithm [11] in order to group millions of sequences into their representative clusters.
>
> — **Prosite** (▶ https://prosite.expasy.org/) uses a sequence classification that is based on **patterns and motifs** that generally are conserved during evolution within in each family. Such patterns are often important for the protein function.
>
> — **InterPro** (▶ http://www.ebi.ac.uk/interpro/) classifies proteins into **families** and predict the presence of domains and important sites. InterPro uses predictive models, known as signatures, in order to provide the classification.
>
> — **Pfam** (▶ http://pfam.xfam.org/) classifies proteins into **domains** [7], which are defined through a statistical model called hidden Markov models (HMMs). From an HMM's perspective, sequences are considered as a succession of events as we move through the sequence. The probability of occurrence of a given amino acid is determined by the observed probabilities in a multiple sequence alignment of the proteins sharing the same function. In that way, regions in the sequence can be identified as functional domains.
>
> — **Gene Ontology (GO)** (▶ http://geneontology.org/) is a highly-rich annotation system, which help classifying protein sequences based on (a) **molecular function**, for instance their biochemical activity; (b) **cellular component**, for instance if the sequence corresponds to a structural protein subunit or in which cellular compartment is localized; (c) **biological process**, for instance if the protein is involved in a catabolic process or some specific pathway of the cell.
>
> — **PANTHER** (▶ http://www.pantherdb.org/) is a classification system based on **evolutionary relationships**.

on homology, such as SWISS-MODEL,[5] MODELLER[6] and many others. Every year, the CASP (Critical Assessment of Techniques for Protein Structure Prediction) competition evaluates the results of structure prediction software from several groups world-wide.[7] The published results and reports of such assessment are useful in order to gain some insights on current progress in various aspects of protein structure prediction.

Structural analysis of enzyme families requires of especialized modeling software such as PyMOL[8] or CHIMERA.[9] The advantage of these tools is that they provide a Python interface, so that they can be potentially integrated with the rest of the analysis that we are performing in this textbook. In our case, what would be of interest is to perform a structural alignment of protein structure in the same enzyme family and see how different variants interact with the substrate. In some cases, comparing 3D structures will reveal interesting features that are not easily detectable by comparing sequences. Many software for structural alignment are available, for instance MAMMOTH [14] and Dali [8].

Returning to our example for PAL from *Petroselinum crispum* (parsley) with UniProt identifier P24481, there is a structure available in the PDB for this protein with identifier 1W27 as shown in ◘ Fig. 5.2. We can now search for similar structures in the PDB by

---

5   ▶ https://swissmodel.expasy.org/
6   ▶ https://salilab.org/modeller/
7   ▶ http://predictioncenter.org/
8   ▶ https://pymol.org
9   ▶ https://www.cgl.ucsf.edu/chimera/

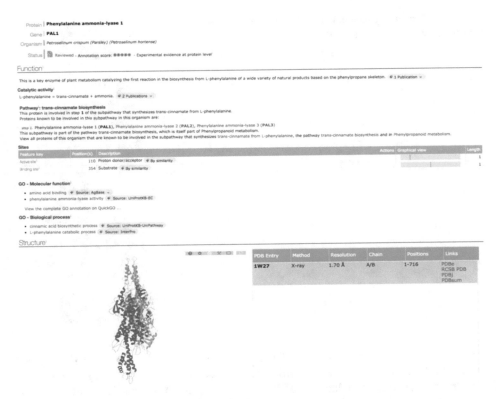

Protein | **Phenylalanine ammonia-lyase 1**

Gene | **PAL1**

Organism | Petroselinum crispum (Parsley) (Petroselinum hortense)

Status | Reviewed - Annotation score: ●●●●● - Experimental evidence at protein level

Function

This is a key enzyme of plant metabolism catalyzing the first reaction in the biosynthesis from L-phenylalanine of a wide variety of natural products based on the phenylpropane skeleton. ● 1 Publication ▾

Catalytic activity
L-phenylalanine = trans-cinnamate + ammonia. ● 2 Publications ▾

Pathway: trans-cinnamate biosynthesis
This protein is involved in step 1 of the subpathway that synthesizes trans-cinnamate from L-phenylalanine.
Proteins known to be involved in this subpathway in this organism are:
  step 1. Phenylalanine ammonia-lyase 1 (**PAL1**), Phenylalanine ammonia-lyase 2 (**PAL2**), Phenylalanine ammonia-lyase 3 (**PAL3**)
This subpathway is part of the pathway trans-cinnamate biosynthesis, which is itself part of Phenylpropanoid metabolism.
View all proteins of this organism that are known to be involved in the subpathway that synthesizes trans-cinnamate from L-phenylalanine, the pathway trans-cinnamate biosynthesis and in Phenylpropanoid metabolism.

Sites

Feature key	Position(x)	Description		Actions	Graphical view	Length
Active site	110	Proton donor/acceptor ● By similarity				1
Binding site	354	Substrate ● By similarity				1

GO - Molecular function
- amino acid binding ● Source: AgBase ▾
- phenylalanine ammonia-lyase activity ● Source: UniProtKB-EC
View the complete GO annotation on QuickGO ...

GO - Biological process
- cinnamic acid biosynthetic process ● Source: UniProtKB-UniPathway
- L-phenylalanine catabolic process ● Source: InterPro

Structure

PDB Entry	Method	Resolution	Chain	Positions	Links
1W27	X-ray	1.70 Å	A/B	1-716	PDBe RCSB PDB PDBj PDBsum

Fig. 5.2 Information about PAL enzyme from *Petroselinum crispum* P24481 in UniProt

performing a structural alignment. Figure 5.3 shows the structure alignment of 1W27 with the top 10 homologues as generated by Dali.

In summary, searching in databases for homologues of some parent enzyme sequence will help us in order to identify other sequences in the same class or family. By mining the wealth of information provided in databases such as structural information, evolutionary features, enzyme activity, etc., we can always narrow down the starting list in the selection of candidate enzyme sequences.

5.2 Enzyme Discovery Through Reaction Homology

If we are interested in finding an enzyme sequence for a desired reaction in the metabolic pathway, our query would be the reaction itself rather than the sequence. We need therefore some mining tool equivalent to BLAST for sequences but for reactions. This can be achieved by computing the **reaction similarity** between the query reaction and the reactions in a database. There exist multiple reaction databases: MetaCyc,[10] KEGG,[11] Rhea,[12]

10 ► https://metacyc.org/
11 ► https://www.genome.jp/kegg/
12 ► https://www.rhea-db.org/

Fig. 5.3   Structural alignment of multiple PAL homologues generated by `Dali`. Image generated using `Chimera`

Reactome,[13] Brenda,[14] SABIO-RK,[15] etc. Each one has some advantages compared with the others. In order to decide which reaction database to use, a solution is to focus on a "meta database", i.e., a database of databases that unifies the identifiers and consensus representations. MetaNetX[16] [13] provides such type of facility. At the time of this writing, MetaNetX contains 42,908 different reactions.

In Code 5.1, we can see an example of downloading the reaction in the Rhea database corresponding to the PAL enzyme database from the website. PAL has the identifier 21385, once downloaded in the example it is converted into an RDKit object.

■ ■ **Code 5.1 Downloading a reaction from Rhea into RDKit.**

```
from rdkit import AllChem
import requests
import tempfile
import os
url = 'https://www.rhea-db.org/rest/1.0/ws/reaction/rxn/21385'
r = requests.get(url)
f = tempfile.NamedTemporaryFile(delete=False)
f.write(r.content)
f.close()
rxn = AllChem.ReactionFromRxnFile(f.name)
os.unlink(f.name)
display( rxn )
```

13 ▶ https://reactome.org/
14 ▶ https://brenda-enzymes.org
15 ▶ http://sabio.h-its.org/
16 ▶ https://www.metanetx.org/

The downloaded PAL reaction can then be easily coded into SMIRKS (SMILES) through RDKit, as shown in Codes 5.2 and 5.3.

■■ Code 5.2 Coding of the chemical reaction using SMIRKS (SMILES).

```
from rdkit import AllChem
smarts = AllChem.ReactionToSmiles(rxn)
print( smarts )
```

■■ Code 5.3 Ouput SMIRKS (SMILES) encoding of the reaction imported in Code 5.2.

```
'[NH3+]C(CC1=CC=CC=C1)C(=O)[O-]>> O=C([O-])/C=C/C1=CC=CC=C1.[H]
 [N+]([H])([H])[H]'
```

The next step is to compute **reaction similarity** between the target reaction and reactions in the database. There are multiple ways of computing reaction similarity. Some tools can help us on this task, for instance Reaction Decoder Tool (RDT)[17] [16] compares bond changes, reaction centers or substructures.

One approach to compute similarity between two chemical species is to employ chemical fingerprints (see also ▶ Sect. 4.2). The steps are as follows:

1. **Fingerprints** are binary vectors where each chemical feature is represented by one or more components of the vector as bits [12]. The features spanned by the vector are generally related to topological elements of the chemical structure like fragments, pharmacophores, etc. Since the number of components that exist in the chemical space can become very large, fingerprint vectors are usually limited in size through a compression technique known as hashing. Hashed fingerprints can be computed by several chemoinformatics toolboxes like RDKit for Python or through workflow systems like KNIME [6].

2. To compute **similarity between chemicals**, we compare their fingerprints. For that purpose, there are several metrics available [12], the **Tanimoto similarity** being one of the most widely used. For binary fingerprints, Tanimoto similarity can be computed as the ratio of common bits divided by the total number of bits in the molecules, making this calculation very efficient.

3. In order to compute **similarity between reactions**, we can at least use two types of strategies based on reactant or reactions fingerprints, as follows:
   a. **Pairwise similarities**:
      i. For each left reactant, compute pairwise similarities to each right reactant and keep the closest one;
      ii. For each right reactant, compute pairwise similarities to each left reactant and keep the closest one;
      iii. Average or maximize both left and right closest similarities in order to obtain an overall similarity between the two reactions.

---

17 ▶ https://github.com/asad/ReactionDecoder

b. **Reaction fingerprint**:
  i. Define an overall reaction fingerprint as the difference between the sum of fingerprints of the products and the sum of reactants. Alternatively, it can be computed by using OR and XOR logical operations instead of computing the sum and difference, respectively. However this last approach has the drawback that it does not make difference between chemical features that are created or that are transformed through the chemical reaction;
  ii. Reaction similarity is computed by comparing the resulting fingerprints;
  iii. Compute similarity between the overall fingerprints;
  iv. An alternative solution for reaction fingerprints is to keep a table with two fingerprint vectors, one consisting on the substrate features that are transformed and another one for the product features and measuring similarity in a similar approach as for reactant fingerprints.

Box 5.3   Reaction Directionality
In the previous description, we assumed a preferred direction for the reactions. When this is not the case, we may compute similarity in both directions in order to keep in the results of the query the closest one to the target reaction. In principle, databases like MetaCyc, KEGG or Brenda [1, 4] may contain information about preferred direction of reactions, reversibility and, increasingly, data about free Gibbs energy. When this information is not available, several strategies are possible in order to assign a preferred direction to the reaction.

The list of candidate sequences obtained in the previous step might contain redundancies as well as overrepresentation for some reaction classes. Therefore, the next step is to analyze the set in order to characterize each protein according to key properties and to ultimately provide some ranking leading to candidate selection.

1. Based on the information contained in the database, reactions can be classified according to their reaction **EC class**.[18] This information can help in order to perform a quick initial pre-selection.
2. Redundancy in sequences can be easily identified through CD-HIT [11], which provides a fast way of clustering large sets of sequences;
3. A multiple sequence alignment of the sequences is then applied to get a more detailed comparative analysis of the sequences with the advantage that it can help to identify conserved regions in the protein, using t-coffee [19] or clustalW[19];
4. Additional position-based and global properties can be computed using packages like EMBOSS [17], amino acid indexes [10] or online tools like PredictProtein[20];
5. Other properties: PSSM (position-specific scoring matrices) using PSI-BLAST, functional domains using Pfam collections and HMMER.[21]

Once the list of sequence candidates with calculated properties has been generated, the list needs to be **prioritized**. The criteria for ranking might vary depending on the application.

---

18 Enzyme Commission number nomenclature (▶ https://enzyme.expasy.org/).
19 ▶ https://www.ebi.ac.uk/Tools/msa/clustalo/
20 ▶ https://www.predictprotein.org/
21 ▶ http://pfam.xfam.org/

For enzyme sequences that are going to be expressed in heterologous host, a ranking can be formed based on the following criteria:

1. **Protein properties**: percentage of secondary structure (helices, sheets, turns), molecular weight, isoelectric point, percentage of polar amino acids, etc;
2. **Functional properties**: target reaction similarity, UniProt protein evidence, sequence conservation in the alignment, etc;
3. **Host-specific properties**: sequence taxonomic distance to host organism, solubility using SCRATCH,[22] protein toxicity using PandaTox.[23]

By combining together these properties through **weighted sums**, we can obtain a score for the sequence. Weights in the score can be heuristically fitted by downloading the proteome sequences of the host organism and calculating typical values obtained for each of these parameters. A more complex scoring function can be also developed by employing **machine learning** using host-specific values for computed protein properties as a training set. We will come back into the application of machine learning for enzyme sequence selection later in ► Chap. 9.

Selenzyme is a free online tool for enzyme selection[24] that integrates the aforementioned features to efficiently search for enzyme sequences starting from some target reaction [3]. Selenzyme works through the following input-output specifications:

1. User queries consist of a **target reaction** expressed in SMILES format or cross-referenced to an external database identifier or EC classification;
2. The software outputs a table of **candidate sequences** that can be ranked based on different criteria. In addition, the user can manually add or remove additional sequences to the table. A multiple sequence alignment of the output table can be visualized through MSAviewer [20].

As an example, we query Selenzyme with the reaction SMARTS that was obtained in Code 5.2. We can perform the query directly on the website or we can use the RESTful web service (see ► Box 5.4) in order to perform the query through a workflow or a script. Code 5.4 uses the requests Python library to communicate with Selenzyme using the JSON data-interchange format. Several properties are computed by Selenzyme for the sequences that are annotated for the closest reactions to the target. A score is calculated for each sequence based on decreasing order by reaction similarity, taxonomic distance, sequence conservation and protein evidence. ◼ Table 5.1 provides a summary for the top 9 predicted sequences.

■■ **Code 5.4 Coding of the chemical reaction using SMIRKS (SMILES).**

```
import json
import pandas as pd
url = "http://selenzyme.synbiochem.co.uk/REST/Query"
data={'smarts':smarts}
response = requests.post(url, json=data)
resp = json.loads( response.text )
tab = pd.DataFrame(json.loads(resp[data']))
```

---

22 ► http://scratch.proteomics.ics.uci.edu/
23 ► http://exploration.weizmann.ac.il/pandatox
24 ► http://selenzyme.synbiochem.co.uk

Box 5.4   Web Services and Workflows

Web services that conform to Representational State Transfer (REST) or RESTful web services, provide interoperability between computer systems on the Internet. Such system has become widespread, especially for cloud computing and mobile applications. The most common operations are through HTTP based on GET and POST and JSON (JavaScript Object Annotation) is the preferred data-interchange format. The adoption of REST by the bioinformatics and systems and synthetic biology communities is allowing a better standardization and a more efficient shared used of the infrastructures. Workflow design tools such as `Galaxy` (▶ https://usegalaxy.org), `KNIME` (▶ http://knime.org/) or `Taverna` (▶ https://taverna.apache.org/) as well as international initiatives such as `Elixir` (▶ https://www.elixir-europe.org/) generally take advantage of RESTful web services in order to perform distributed operations.

▣ Table 5.1   Top enzyme sequences selected by `Selenzyme`

% helices	% sheets	Organism source	Polar %	Score	Seq. ID	Tax. distance
47.7	22.7	*Taxus wallichiana*	43.81	75.9	Q68G84	24
43.2	18.7	*Persea americana*	45.97	75.8	P45727	24
48.5	18.7	*Pinus taeda*	46.02	75.7	P52777	24
46.0	18.5	*Camellia sinensis*	45.66	72.8	P45726	27
46.8	15.8	*Vitis vinifera*	49.04	71.8	P45735	28
46.3	19.9	*Bromheadia finlaysoniana*	45.09	71.8	Q42609	28
43.2	19.6	*Arabidopsis thaliana*	46.34	70.9	P35510	29
45.0	21.2	*Arabidopsis thaliana*	45.96	70.9	P45725	29
46.7	19.8	*Arabidopsis thaliana*	45.69	70.9	Q9SS45	29

5.3   Enzyme Discovery in the Extended Metabolic Space

In the previous sections we had an overview about techniques allowing the screening of databases in order to identify close homologues in the known sequence space through sequence, structure or reaction similarity. Our basic assumption was that current biochemical knowledge should be able to provide an answer to the question of finding a sequence for an enzyme with some predefined properties. Obviously, a more ambitious goal would be to interrogate known biochemical databases in a prospective way about the extent of **hypothetical enzymes that are likely candidates for some target activity**. In other words, can our current knowledge help us in order to model an **extended metabolic space** that considers any chemical that potentially could be produced through biosynthetic pathways?

In ▶ Chap. 2 we learned about genome-scale metabolic models of organisms. Nowadays, accurate models of metabolism exist for a large number of organisms across the tree of life. Each organism is defined by a set of metabolic reactions catalyzed by enzymes which will lead to the production of a defined set of metabolites. The set of metabolites that are specific to a given organism defines its metabolome (see ▶ Box 5.5).

The metabolome of an organism, however, is highly dynamic and this is why we study the metabolic phenotypes of organisms as they respond to stimuli or vary through time. For instance, analyzing the human metabolic phenotype has become a promising way of early detection of diseases providing novel ways for performing diagnostics [9].

Box 5.5 "Omics" and "Omes"
- **Genomics** analyzes the genome of an organism, i.e. their complete set of DNA sequences.
- **Transcriptomics** analyzes the transcriptome, i.e., the set of mRNA sequences expressed from the genes in an organism at some experimental conditions.
- **Proteomics** quantifies the concentrations of proteins in the proteome, i.e., the set of proteins in an organism.
- **Metabolomics** quantifies the concentrations of chemical compounds in the metabolome of an organism.
- **Reactomics** characterizes the set of biochemcial reactions in an organism, i.e. its reactome.

Ideally, we should be able to expand the capabilities of metabolic models for enzyme discovery by considering **de novo reactions**. As already discussed in ▶ Chap. 4, enzymes have the ability of catalyzing multiple reactions and to accept multiple substrates. In nature, enzymes show specialization towards some catalytic activities because these activities are the ones that they have evolved through evolutionary pressure. However, other latent transformations are generally possible to be catalyzed by the enzyme. This observation has been used in biotechnology as a way for exploring ways of synthetizing novel compounds. For instance, in order to synthetize a drug that is an analog of some other with some target bioactivity.

Let's start by focusing on the similarities and differences between organisms by looking at the metabolomes as defined by their respective metabolic models. A basic assumption that we can make here is that all metabolites contained in the model are actually present in the cell. In that way a quick comparison between the different models can be performed simply by looking at their metabolites. For instance, we can compare the models of *E. coli* and *Salmonella* that are in the `cobrapy` toolbox by running Code 5.5. ◻ Table 5.2 provides the resulting differences. Both models share approximately 80% of the total metabolites ($\approx 2,000$), while the other 20% is almost equally distributed for each model.

■ ■ Code 5.5 Metabolome comparison between the *E. coli* and the *Salmonella* models.

```
import cobra.test
eco = cobra.test.create_test_model("ecoli")
salm = cobra.test.create_test_model("salmonella")
em = [x.id for x in eco.metabolites]
sm = [ x.id for x in salm.metabolites]
total = set(sm) | set(em)
common = set(sm) & set(em)
unieco = set(em) - set(sm)
unisam = set(sm) - set(em)
print(total, common, unieco, unisam)
```

■ **Table 5.2** Comparison between *E. coli* and *Salmonella* metabolomes

Set	Number	Percentage
Total	2014	100%
Common	1593	79.1%
Unique *Salmonella*	212	10.5%
Unique *E. coli*	209	10.4%

As inferred from ■ Table 5.2, each organism hosts its own defined chemical space determining in that way a chemical tree of life (see ▶ Box 5.6) which is comprised by the full set of metabolites that are catabolized and broken down. However, such chemical space defined by the models is just a subset of the full set of metabolites that the organism could potentially produce (or degrade) due to the substrate versatility of enzymes. We define the **extended metabolic space** as the chemical space that contains not only the chemicals that are generally produced by enzymes in metabolic networks, but also those that could be potentially produced through promiscuity.

> **Box 5.6   The Chemical Tree of Life**
> DNA sequencing technologies allow generating phylogenetic trees defining the taxonomy of organisms based on consensus sequences that appear across the species. Several initiatives exist to generate a consensus Tree of Life reflecting the past evolutionary history from genomics information, such as iTOL (▶ http://tolweb.org). Interestingly, another approach exists based on classifying the organisms based on their metabolic phenotypes. Strikingly, a tree of life based on sequence information and a tree of life based on chemical information share close similarity, illustrating the fact that both enzyme sequences and chemical reactions in organisms are intimately related and are the footprints of their evolutionary history.

Estimating the extended metabolic space might seem a challenging task. Nevertheless, performing an initial estimate is possible by modeling enzyme promiscuity through **generalized reaction rules** as introduced in ▶ Sect. 4.3. What is required in order to estimate the extended metabolic space would be a complete database of reaction rules that covers the full set of known chemical transformations that can potentially be performed through biocatalysis. RetroRules[25] [5] is a database of reaction rules for the full set of known reactions in the biochemical space that contains representations of the rules at different levels of specificity. Starting with such database, the generation of an extended metabolic space is possible. The steps in order to generate the extended metabolic space are similar to the ones that we used in ▶ Sect. 4.4 for enumerating chemical reactions with the difference that we will apply them now to all known reactions for some metabolic network, such as those contained in the models for *E. coli* or *Salmonella*.

---

25 ▶ https://retrorules.org/

The steps in order to generate the extended metabolic space are as follows:

1. Start with a designated set of **reactions of interest**, for instance the ones that are specific to some organism, some pathway, or in general the ones that are contained in metabolic databases;
2. Determine the **initial set of metabolites** such as the ones that are present in the model of the organism;
3. **Map each reaction** into its corresponding reaction rules using `RetroRules`;
4. For each reaction rule, **enumerate all possible products** by running the reaction through the set of metabolites in the same way as was performed in Code 4.13;
5. Add the new products to the set of metabolites and **repeat the enumeration procedure** as many times as desired or until convergence is reached.

The previous procedure can potentially generate a large number of new metabolites, especially if reaction rules of low specificity are allowed in the enumeration. Therefore, we need to add some cutoff such as limiting the molecular weight of the product (in order to avoid endless polymerization) or limiting the total number of products generated by a single reaction (in order to avoid highly generative reactions).

At the end of the enumeration procedure, the total set of metabolites $\hat{\mathbf{x}}$ will consist of the initial metabolite species $\mathbf{x}$ augmented by the de novo metabolites $\mathbf{x}_e$ predicted by the enumeration procedure. The **stoichiometric matrix** of the model as defined in Eq. 2.4 is augmented through the extended metabolic space so that the change in time of the concentration of total metabolites $\hat{\mathbf{x}}$ becomes as follows:

$$\dot{\hat{\mathbf{x}}} = \begin{bmatrix} \dot{\mathbf{x}} \\ \dot{\mathbf{x}}_e \end{bmatrix} = \begin{bmatrix} \mathbf{S} & \mathbf{S}_{xe} \\ \mathbf{S}_{ex} & \mathbf{S}_e \end{bmatrix} \begin{bmatrix} \mathbf{v} \\ \mathbf{v}_e \end{bmatrix}. \tag{5.1}$$

This equation resembles Eq. 3.1, with the fundamental difference that whether in Eq. 3.1 we augmented the stoichiometric matrix of a host organism by adding the reactions of an engineered pathway; in Eq. 5.1 we added the expression for reactions that potentially could occur in the cell (according to our model of enzyme promiscuity) as long as the right precursors are present in the media.

We would work out now an **example of calculation of the extended metabolic space for** *E. coli*. In order to do that the first thing that we need are the chemical structures of its metabolome. So far, we have been using the *E. coli* model that comes by default in the `cobrapy` package. However, that model does not contain direct information about the chemical structures, at least at the time of this writing. This lack of information might be explained because, originally, genome-scale models were mainly focused on the computation of network-wide properties such as flux balance analysis, pathways or topological clusters, and chemical structures were less relevant. As we have learned in this textbook, such trend is now moving into integrating the information about the chemical and sequence space as part of the analysis of genome-scale models. Therefore, recent versions of genome-scale models provide more detailed annotations for chemical and reaction information.

In our example, we will consider *E. coli* model `iAF1260`, which can be downloaded from the `BioModels` database[26] that includes cross-references to other databases facili-

---

26 ▶ http://identifiers.org/biomodels.db/MODEL3023609334

tating our present analysis. The SBML model will be downloaded into data/iAF1260.
xml and the cross-reference file from the MetaNetX[27] consensus database [13] will be
downloaded into metanetx/chem_prop.tsv. We proceed now with Code 5.6 in
order to store each metabolite as an RDKit mol object in the ecomol set and its associated InChI code into ecoinchi. Similarly, the reactions present in the model will be
stored into ecor using their MetaNetX identifiers.

■ ■ **Code 5.6 Extraction of chemical structures and reactions of the** *E. coli* iAF1260
metabolic model.

```python
import cobra
from rdkit import Chem
def readChem(f):
    chm = {}
    with open(f) as h:
        for line in h:
            if line.startswith('#'):
                continue
            row = line.rstrip().split('\t')
            cid = row[0]

            chm[cid] = row
    return chm
chm = readChem('metanetx/chem_prop.tsv')
eco2 = cobra.io.read_sbml_model('data/iAF1260.xml')
ecomol = set()
ecoinchi = set()
mm = [m for m in eco2.metabolites]
for m in mm:
    if 'metanetx.chemical' in m.annotation:
        cid = m.annotation['metanetx.chemical']
        if cid in chm:
            mol = Chem.MolFromSmiles( chm[cid][6] )
            if mol is not None:
                ecomol.add( mol )
                ecoinchi.add( Chem.inchi.MolToInchi(mol) )
ecor = set()
for r in eco2.reactions:
    if 'metanetx.reaction' in r.annotation:
        ecor.add(r.annotation['metanetx.reaction'])
```

The set of pre-computed reaction rules in RetroRules[28] has to be then downloaded as
shown in Code 5.7. RetroRules provides sets of pre-computed forward and reversed
reactions (see ▶ Box 5.7). In the next example, we will focus on the forward rules as these
are the ones that should allow the expansion of the metabolic space. For our exercise, we
will focus on reactions that are described at diameter = 12. Remember from ▶ Sect. 4.3
that the diameter of a reaction rules allows representing the degree of promiscuity in the
reaction, i.e., larger diameters belong to higher specific reactions whereas low diameters
correspond with generalist reactions. In Code 5.7 the RetroRules reaction rules of

---

27 ▶ https://www.metanetx.org/
28 ▶ https://retrorules.org/

diameter = 12 that are present in the *E. coli* model (in the `ecor` set) are converted into `RDKit` reaction objects. Note that in Code 5.7 we have used the cross-link through the `MetaNetX` identifier in order to determine if the reaction is present in the *E. coli* model.

Box 5.7
`RetroRules` provides reaction rules expressed in the SMARTS format (see ▶ Sect. 4.2) with an augmented presentation at different levels of specificity (the atomic environment around the reaction center). In order to simplify the use of reaction rules in enumeration procedures, `RetroRules` are based on transformations focused on a single reactant.

For instance a reaction such as $A + B \rightarrow C + D$ becomes divided into 4 different rules based on their corresponding atom-atom mappings (see mapping examples in ◘ Figs. 4.5 and 4.7):

$$
\begin{aligned}
R_1: \quad & A \rightarrow C + D, \\
R_2: \quad & B \rightarrow C + D, \\
R_3: \quad A + B \leftarrow & C, \\
R_4: \quad A + B \leftarrow & D.
\end{aligned}
\tag{5.2}
$$

The reason for using forward (R_1 and R_2) and reverse (R_3 and R_4) representations is because this allows for both metabolic expansion and retrosynthesis analysis, which will described later in ▶ Chap. 6.

Such single reactant representation has some implications in the case of computing a metabolic space like the extended metabolic space. If a reaction involves more than one reaction (apart from the common co-factors, co-substrates, etc.), it would be necessary to verify that all reactants are actually present in the initial set of metabolites.

■■ **Code 5.7 Collection of reaction rules of diameter 12 from** `RetroRules`.

```
rf = 'retrorules_preparsed/retrorules-rr01_flat_forward.csv'
rl = []
with open(rf) as h:
    cv = csv.DictReader(h)
    for row in cv:
        diam = row['Diameter']
        if row['Diameter'] == '12':
            ruid = row['Rule_ID']
            rid, cid = ruid.split('_')
            rl.append( rid )
            continue
        if rid in ecor:
            rxn = AllChem.ReactionFromSmarts( row['Rule'] )
            rl.append( rxn )
```

Finally, we perform in Code 5.8 a basic procedure that will run **a first iteration of expansion of the metabolic space** of *E. coli* by predicting new metabolites present according to the set of reaction rules of diameter = 12. In this example, we obtained a total expansion of 1% of the *E coli* metabolome. This is a rather modest expansion because we have tried to keep the example computationally simple. Much larger extended metabolic spaces can be obtained by relaxing the constraints as follows:

1. Increasing the number of iterations;
2. Lowering the diameter of the reaction rule;
3. Considering also additional metabolites that can be in the media.[29]

■■ **Code 5.8 Metabolic space expansion for the reaction rules of diameter 12 from** `RetroRules`.

```
ecoems  = set ( [ i for i in ecomol ] )
ecoinchiems = set([i  for  i  in  ecoinchi])
for  r  in  rl :
    for  y  in  ecomol :
        try :
            pr = r.RunReactants ( (y,)  )
            for  s  in  pr :
                inchi = Chem.inchi.MolToInchi( s[0]  )
                if  inchi  not  in  ecoinchiems :
                    ecoems.add(inchi)
                    ecoinchiems.add(inchi)
                    new.add(inchi)
        except :
            continue
```

Take Home Message

— Enzyme discovery and selection allows searching for best candidate sequences catalyzing the target reactions in the pathway.
— Enzyme discovery through sequence homology involves analyzing a multiple sequence alignment of the enzymes in the same family.
— Another approach to enzyme discovery when no sequence is known for the target reactions consists on searching for similar chemical reactions with annotated sequences.
— The extended metabolic space contains putative reactions based on enzyme promiscuity and can be used to expand pathway chemical diversity.

5.4 **Problems**

? **5.1** Enzyme selection based on experimental efficiency.

✓ Start with transaminase enzyme activity and determine its reference EC number.
✓ Go to Brenda, look for the sequence with higher affinity to the substrate.
✓ Go to Brenda, look for the sequence with higher turnover rate.
✓ Look for the sequence with the highest efficiency.
✓ Which sequence would you select?

---

29 The E. coli metabolome database (ECMDB) ▶ http://ecmdb.ca/, similarly to HMDB for human as discussed in ▶ Chap. 4, provides a large set of metabolites observed in *E. coli* from metabolomic data.

? **5.2** Enzyme selection based on phylogenetic closeness.

✓ Go to `MetaCyc`, find a sequence that is provided as template for the transaminase given reaction.

✓ Perform a BLAST of the sequence, look for the homologue that is closer to *E. coli*.

? **5.3** Enzyme selection based on sequence similarity.

✓ Go to `MetaCyc`, find a sequence that is provided as template for the transaminase given reaction.

✓ Perform a BLAST of the sequence and keep the top 100 hits.

✓ Align the sequences (using UniProt).

✓ Draw a phylogenetic tree of the results using `iTOL` (see ▶ Box 5.6).

✓ Highlight those sequences that are annotated with the target activity.

✓ Identify at least 5 sequences that are not annotated with the target activity but appear close in the phylogenetic tree.

✓ Based on their annotated actitivies, what are the chances that these homologues will also functional for the target activity?

? **5.4** Extended metabolic space (EMS) of the full *E. coli* metabolome.

✓ Use `MetaCyc` in order to download the metabolome with their InChI instead of using the *E. coli* model.

✓ Repeat the EMS procedure, do you see some differences?

? **5.5** Extended metabolic space (EMS) of *E. coli* at the lowest specificity.

✓ Perform the same *E. coli* EMS procedure as described in ▶ Sect. 5.3 but at the lowest diameter, i.e., a diameter = 2.

✓ How many new metabolites do you obtain after the first and second iteration?

? **5.6** Extended metabolic space (EMS) of the reachable space of *E. coli*.

✓ Perform the same *E. coli* EMS procedure as described in ▶ Sect. 5.3 now for the full set of reactions at some diameter.

✓ In this case, what you are calculating is the extended reachable space of *E. coli*.

✓ How many new metabolites do you obtain after the first and second iteration?

? **5.7** Computing the extended biochemical space (EMS).

✓ Perform the same EMS procedure as described in ▶ Sect. 5.3 now for the full set of reactions and full set of metabolites (from `MetaCyc` or from `MetaNetx` (this one is much larger)) at some diameter.

✓ In this case, what you are calculating is the extended full biochemical space.

✓ How many new metabolites do you obtain after the first and second iteration?

References

1. Altman, T., Travers, M., Kothari, A., Caspi, R., Karp, P.D.: A systematic comparison of the MetaCyc and KEGG pathway databases. BMC Bioinformatics **14**(1), 112 (2013). https://doi.org/10.1186/1471-2105-14-112
2. Arnold, F.H.: Enzymes by evolution: bringing new chemistry to life. Mol. Front. J. **2**(01), 9–18 (2018). https://doi.org/10.1142/S2529732518400023
3. Carbonell, P., Wong, J., Swainston, N., Takano, E., Turner, N.J., Scrutton, N.S., Kell, D.B., Breitling, R., Faulon, J.L.: Selenzyme: enzyme selection tool for pathway design. Bioinformatics **34**(12), 2153–2154 (2018). https://doi.org/10.1093/bioinformatics/bty065
4. Chang, A., Schomburg, I., Placzek, S., Jeske, L., Ulbrich, M., Xiao, M., Sensen, C.W., Schomburg, D.: BRENDA in 2015: exciting developments in its 25th year of existence. Nucl. Acids Res. gku1068 (2014). https://doi.org/10.1093/nar/gku1068
5. Duigou, T., du Lac, M., Carbonell, P., Faulon, J.L.: RetroRules: a database of reaction rules for engineering biology. Nucl. Acids Res. (2018). https://doi.org/10.1093/nar/gky940
6. Fillbrunn, A., Dietz, C., Pfeuffer, J., Rahn, R., Landrum, G.A., Berthold, M.R.: KNIME for reproducible cross-domain analysis of life science data. J. Biotechnol. (2017). https://doi.org/10.1016/j.jbiotec.2017.07.028
7. Finn, R.D., Coggill, P., Eberhardt, R.Y., Eddy, S.R., Mistry, J., Mitchell, A.A., Potter, S.C., Punta, M., Qureshi, M., Sangrador-Vegas, A., Salazar, G.A., Tate, J., Bateman, A.: The Pfam protein families database: towards a more sustainable future. Nucl. Acids Res. **44**(D1), D279–D285 (2016). https://doi.org/10.1093/nar/gkv1344
8. Holm, L., Laakso, L.M.: Dali server update. Nucl. Acids Res. **44**(W1), W351–W355 (2016). https://doi.org/10.1093/nar/gkw357
9. Johnson, C.H., Ivanisevic, J., Siuzdak, G.: Metabolomics: beyond biomarkers and towards mechanisms. Nat. Rev. Mol. Cell Biol. **17**(7), 451–459 (2016). https://doi.org/10.1038/nrm.2016.25
10. Kawashima, S., Kanehisa, M.: AAindex: amino acid index database. Nucl. Acids Res. **28**(1), 374 (2000). https://doi.org/10.1093/nar/28.1.374
11. Li, W., Godzik, A.: Cd-hit: a fast program for clustering and comparing large sets of protein or nucleotide sequences. Bioinformatics **22**(13), 1658–1659 (2006)
12. Maggiora, G.M., Shanmugasundaram, V.: Molecular similarity measures. Methods Mol. Biol. (Clifton, N.J.) **672**, 39–100 (2011). https://doi.org/10.1007/978-1-60761-839-3_2
13. Moretti, S., Martin, O., Tran, T.V.D., Bridge, A., Morgat, A., Pagni, M.: MetaNetX/MNXref reconciliation of metabolites and biochemical reactions to bring together genome-scale metabolic networks. Nucl. Acids Res. **44**(D1), D523–D526 (2016). https://doi.org/10.1093/nar/gkv1117
14. Ortiz, A.R., Strauss, C.E., Olmea, O.: MAMMOTH (matching molecular models obtained from theory): an automated method for model comparison. Protein Sci. **11**(11), 2606–2621 (2002). https://doi.org/10.1110/ps.0215902
15. Prather, K.L.J.: Chemistry as biology by design. Microbial Biotechnol. (2018). https://doi.org/10.1111/1751-7915.13345
16. Rahman, S.A., Torrance, G., Baldacci, L., Martínez Cuesta, S., Fenninger, F., Gopal, N., Choudhary, S., May, J.W., Holliday, G.L., Steinbeck, C., Thornton, J.M.: Reaction Decoder Tool (RDT): extracting features from chemical reactions. Bioinformatics **32**(13), 2065–2066 (2016). https://doi.org/10.1093/bioinformatics/btw096
17. Rice, P., Longden, I., Bleasby, A.: EMBOSS: the European molecular biology open software suite. Trends Genet. **16**(6), 276–277 (2000)
18. Suzek, B.E., Wang, Y., Huang, H., McGarvey, P.B., Wu, C.H.: UniRef clusters: a comprehensive and scalable alternative for improving sequence similarity searches. Bioinformatics **31**(6), 926–932 (2015). https://doi.org/10.1093/bioinformatics/btu739
19. Taly, J.F., Magis, C., Bussotti, G., Chang, J.M., Tommaso, P.D., Erb, I., Espinosa-Carrasco, J., Kemena, C., Notredame, C.: Using the T-Coffee package to build multiple sequence alignments of protein, RNA, DNA sequences and 3D structures. Nat. Protoc. **6**(11), 1669–1682 (2011). https://doi.org/10.1038/nprot.2011.393
20. Yachdav, G., Wilzbach, S., Rauscher, B., Sheridan, R., Sillitoe, I., Procter, J., Lewis, S.E., Rost, B., Goldberg, T.: MSAViewer: interactive JavaScript visualization of multiple sequence alignments. Bioinformatics **32**(22), 3501–3503 (2016). https://doi.org/10.1093/bioinformatics/btw474

Further Reading

A good introduction to **sequence similarity** searching:
Pearson, W.R.: An Introduction to Sequence Similarity (Homology) Searching. Current protocols in bioin-
formatics/editoral board, Andreas D. Baxevanis … [et al.] (2013).

A good introduction to **protein structure analysis** using `PyMOL`:
Skern, T.: Exploring Protein Structure: Principles and Practice. Learning Materials in Biosciences. Springer
International Publishing, Cham (2018)

EC-BLAST provides a good example of a tool for **enzyme discovery and selection** focused on the reaction:
Rahman, S.A., Cuesta, S.M., Furnham, N., Holliday, G.L., Thornton, J.M.: EC-BLAST: a tool to automatically
search and compare enzyme reactions. Nat. Methods **11**(2), 171–174 (2014)

A formal introduction to **extended metabolic space modeling**:
Carbonell, P., Delépine, B., Faulon, J.L.: Extended metabolic space modeling. In: Methods in Molecular
Biology, vol. 1671, pp. 83–96 (2018)

Pathway Discovery

© Springer Nature Switzerland AG 2019
P. Carbonell, *Metabolic Pathway Design*, Learning Materials in Biosciences,
https://doi.org/10.1007/978-3-030-29865-4_6

What You Will Learn in This Chapter

Pathway discovery is a prospective task that is part of the metabolic pathway design work-flow. Using the tools that were described in previous chapters to model metabolic networks and chemical diversity, we can now start exploring the metabolic space for routes leading to the production of promising targets. Chemical targets of interest can be identified by performing techno-economic and life cycle analyses. A bioretrosynthesis analysis assesses the existence and feasibility of biosynthetic pathways connecting the target to the chassis. Generalized reaction rules in the bioretrosynthesis analysis can be applied in order to predict *de novo* or hypothetical routes based on enzyme promiscuity.

6.1 Defining Chemical Targets

One essential (but not always obvious) question that we need to answer on any metabolic pathway design project is what is actually the goal of our pathway design project. In this chapter, we will look into the different types of goals that can be associated with a cell that is modified by engineering a metabolic pathway in order to start producing (or overproducing) a desired chemical compound. Our target will be therefore some **chemical compound of interest**. We will assume during the rest of the chapter that our interest is to produce some natural compounds, like those generally found in plants that are difficult to harvest or involve costly extraction. When possible, the chemical industry has replaced the production of such type of compounds by synthetic chemistry methods. Thanks to metabolic engineering and synthetic biology, an alternative, potentially more sustainable and greener way of producing the chemical can be available through engineered microbial production.

Major classes of **natural products** of interest include alkaloids, flavonoids, terpenoids, polyketides and so many more. Such families contain thousands of natural compounds that are derived from some initial hub compounds through secondary metabolism. We are going to focus now on discovering pathways producing some of these **hub compounds** since having engineered strains producing them will be the first step in order to be able to produce derivatized compounds of high value in the compound family.

Choosing the chemical target is a decision that involves considering many different aspects. Techno-economic and life cycle analyses, similar to the ones applied in chemical engineering, are useful in order to initially select promising target compounds (see ▶ Box 6.1). However, one of the major difficulties is that it is not always easy to have a precise model a priori of the performance of the process. We can only roughly estimate what will be the **theoretical maximum achievable production yields** of an engineered strain with the optimized pathway. Such problem will be later addressed, but for the moment in this chapter, we will focus on identifying all available production pathways for a desired chemical target. Later in ▶ Sect. 7.2 some approaches using genome-scale models for estimating pathway performance will be discussed.

We present an example of a **decision process for identifying a good target**. As already discussed, target selection is a complex task and therefore what we will show here is going to be a simplified case. Our basic criteria is that we want to identify a good target displaying some specific properties. A good starting point in order to identify such target is

Box 6.1
Techno-economic analysis (TEA) and life cycle analysis (LCA) are tools that help on assessing the overall impact of research projects:
— TEA assesses technical and economic feasibility of the process:
  — Methods are generally based on analyzing the flows involved in the process;
  — Starting from the initial feedstock, maximum expected conversion yields are calculated to the final product, including separation and extraction, based on a model of the process;
  — Process economics considers cash flow analysis with capital and operating costs;
  — As a result, a minimum selling price for the product is estimated.
— LCA assesses the overall environmental impact of the technology[1]:
  — Comparison of different pretreatment processes on the basis of cost, energy, and life-cycle greenhouse gas emissions;
  — Quantify the potential impact of alternative types of feedstocks, i.e., salt-tolerant, draught-tolerant, etc.

◻ **Table 6.1**   Target properties

Class	Constraint	Source
Chemical Role	Antioxidant	CHEBI:22586
Application	Antineoplastic agent	CHEBI:35610
Application	Neuroprotective agent	CHEBI:63726
Druglikeness	Lipinski's rule of five	Hydrogen bond donors ≤ 5 Hydrogen bond acceptors ≤ 10 Molecular mass ≤ 500 daltons Octanol-water partition coefficient ≤ 5

searching in the `ChEBI` database [5]. `ChEBI` is a database of chemical entities of biological interest that is especially valuable due to its ontology classification. Ontologies help organizing our knowledge about natural compounds. More precisely, we can select compounds having some chemical role or because are used for some therapeutic application. In our example, we are going to select compounds with the properties shown in ◻ Table 6.1. Our ideal target will be an antioxidant compound that is an antineoplastic as well as neuroprotective agent. Such type of targets could be prospective candidate drugs for chemotherapy. Moreover, we want to evaluate the druglikeness of the candidate compounds, i.e., if the compound displays chemical and physical properties that make it a likely orally active drug in humans. To that end, we can use the *Lipinski's rule of five* [6]

---

1   LCA is one of the requirements in the International standard ISO 14001:2015

shown in ▣ Table 6.1, which are some practical rules that are often used in the pharmacological industry to quickly evaluate druglikeness.

As shown in ▣ Table 6.1, each ChEBI ontology has a corresponding identifier. For instance, antioxidant activity is represented by ChEBI id = 22586 (*a substance that opposes oxidation or inhibits reactions brought about by dioxygen or peroxides*).

The procedure consists of the following steps:

1. Download the list of antioxidant compounds:
   a. Go to the ChEBI website[2] and search for the term "antioxidant" or directly for its identifier CHEBI:22586, follow the resulting link to will arrive at the ChEBI page for antioxidants;
   b. On the antioxidants page, go to the tab containing the ChEBI ontologies[3];
   c. Download the list of compounds showing antioxidant activity (459 compounds at the moment of this writing) in SDF format by following the provided link.

2. Download the list of antineoplastic agents:
   a. Go to the ChEBI website and search for the term "antineoplastic" or directly for its identifier CHEBI:35610, follow the resulting link to will arrive at the ChEBI page for antineoplastics;
   b. On the antineoplastics page, go to the tab containing the ChEBI ontologies[4];
   c. Download the list of compounds showing antineoplastic activity (1487 compounds at the moment of this writing) in SDF format by following the provided link.

3. Download the list of neuroprotective agents:
   a. Go to the ChEBI website and search for the term "neuroprotective" or directly for its identifier CHEBI:63726, follow the resulting link to will arrive at the ChEBI page for neuroprotective agents;
   b. On the neuroprotective agents page, go to the tab containing the ChEBI ontologies[5];
   c. Download as the list of compounds showing neuroprotective activity (125 compounds at the moment of this writing) in SDF format by following the provided link.

The next step is to select the targets sharing the three aforementioned properties. To that end, we just need to look for those compounds that are shared by the three downloaded groups. In Code 6.1, compounds from each group are read and the common ones are stored in targets. After this selection, there are in total 12 target candidates identified, which will be further reduced to the 7 targets shown in ▣ Table 6.2 in the following.

---

2 ▶ https://wwww.ebi.ac.uk
3 ▶ https://www.ebi.ac.uk/chebi/chebiOntology.do?chebiId=CHEBI:22586
4 ▶ https://www.ebi.ac.uk/chebi/chebiOntology.do?chebiId=CHEBI:35610
5 ▶ https://www.ebi.ac.uk/chebi/chebiOntology.do?chebiId=CHEBI:63726

▫ Table 6.2 Selected targets

Structure	Name	H donors	H acceptors	Mass	logP	Price (10 mg)
	Pyrrolidine dithiocarbamate (PDTC)	1.0	1.0	147.268	1.2969	84
	Pinocembrin	2.0	4.0	256.257	2.8043	168
	Phenethyl caffeate	2.0	4.0	284.311	2.8969	25
	Auraptene	0.0	3.0	298.382	4.8645	140

(continued)

6

◻ Table 6.2 (continued)

Structure	Name	H donors	H acceptors	Mass	logP	Price (10 mg)
	Morin	5.0	7.0	302.238	1.9880	25
	Oleocanthal	1.0	5.0	304.342	2.2184	25
	Curcumin	2.0	6.0	368.385	3.3699	25

■■ **Code 6.1 Select compound targets based on their properties.**

```
def sdfMol(f):
    suppl = Chem.rdmolfiles.SDMolSupplier( f )
    x = {}
    y = set()
    for mol in suppl:
        try:
            chebi = mol.GetProp('NAME')
            y.add( mol.GetProp('NAME') )
            x[chebi] = mol
        except:
            continue
    return x,y
aox, aoxid = sdfMol('data/ChEBI_Antioxidant.sdf')
neo, neoid = sdfMol('data/ChEBI_AntiNeoplastic.sdf')
neu, neuid = sdfMol('data/ChEBI_NeuroProtective.sdf')
targets = []
for cid in aoxid & neoid & neuid:
    targets.append(neo[cid])
```

We first calculate the molecular properties of the target compounds (Code 6.2).

■■ **Code 6.2 Calculation of molecular properties of Lipinski's rules.**

```
import numpy as np
from rdkit.Chem import Descriptors
X = []
for m in targets:
    try:
        v = []
        v.append( Descriptors.NumHDonors( m ) )
        v.append( Descriptors.NumHAcceptors( m ) )
        v.append( Descriptors.MolWt( m ) )
        v.append( Descriptors.MolLogP( m ) )
        X.append(np.array(v))
    except:
        continue
X = np.array( X )
```

Lipinski's rules are only verified by 7 of the 12 candidates, as calculated in Code 6.3.

■ ■ **Code 6.3 Selection of targets based on Lipinski's rules.**

```
ix= np.logical_and.reduce( [
        X[:,0] <= 5,
        X[:,1] <= 10,
        X[:,2] < 500.0,
        X[:,3] < 5
        ] )
newTargets = []
sumTable = []
for i in sorted(np.where(ix)[0], key=lambda j: X[j,2]):
    try:
        print( X[i,2], targets[i].GetProp('NAME') )
        display( targets[i] )
        newTargets.append( targets[i] )
        sumTable.append( [targets[i].GetProp('NAME')]
                + list(X[i,:]) )
    except:
        continue
```

In order to further select a target from these candidates, we can look at the market price of 10 mg of each compound[6] (see ▢ Table 6.2). Our first candidate based on all previous considerations and current market price is the **flavonoid pinocembrin**, which we will use as our target compound for the rest of this chapter.

6.2 Retrosynthetic Analysis of the Metabolic Scope

Once we have identified the target, we need to explore **which biosynthetic routes are available** by looking at both natural metabolic pathways and putative or de novo routes. A first step is to **look into the literature and available databases** such as MetaCyc and KEGG in order to find out if a natural pathway for producing the target compound is known and if all the enzymatic steps are well characterized. Even if the actual pathway for the target compound is not fully known, still will be useful looking at the route that is common to all the compounds in the same class and their natural precursors. ▢ Table 6.3 provides some of such information retrieved from databases and literature for the compounds in the target list.

Besides all the information to be found in pathway databases and in the literature, we can also **manually search for possible routes in reaction databases**. Such strategy would involve starting from the target compound and then to look for all possible enzymatic reactions that can produce the compound. After that initial step, we repeat the process for each precursor as many times as needed until we finally connect the target compound to precursors that are available in the media or as metabolites in the host organism or chassis. Such type of approach is called **retrosynthesis**, a technique that originally was developed for synthetic chemistry. If we restrict such backward reaction search to enzymatic reactions then the process is generally known as **bioretrosynthesis**.

---

6   Verified at ▶ https://www.aldrichmarketselect.com/

Table 6.3 Prospective search for metabolic pathways

Compound	Pathway	Sources
PDTC	Not available (N.A.)	Chemical synthesis
Pinocembrin	Flavonoid	Plants and propolis
Phenethyl caffeate	N.A.	Propolis and essential oils
Auraptene	N.A.	Citrus
Morin	N.A.	Fruits and vegetables
Oleocanthal	Flavonol	Olive oil
Curcumin	Polyketide	Turmeric

Table 6.4 Retrosynthetic search for pinocembrin

Step	Enzyme	Product	Substrate	EC number
1	CHI	(2S)-pinocembrin	Pinocembrin chalcone	5.5.1.6
2	CHS	Pinocembrin chalcone	(E)-cinnamoyl-CoA	2.3.1.-
3	4CL	(E)-cinnamoyl-CoA	Trans-cinnamate	6.2.1.1
4	PAL	Trans-cinnamate	L-phenylalanine	4.3.1.24

For the target pinocembrin and in general for any other chemical target we can go to online metabolic databases such as KEGG or MetaCyc in order to start building the retrosynthetic map. For instance, Metacyc lists all reactions that are known to produce the compound.[7] In our case, the only reaction that appears in the database producing pinocembrin is a chalcone isomerase (CHI), which produces pinocembrin from pinocembrin chalcone. If pinocembrin chalcone is not in our chassis as this is the case for many common chassis used in metabolic engineering like *Escherichia coli* or *Saccharomyces cerevisiae*, then we need to iteratively apply the search in order to move upstream in the pathway until we would hopefully find a substrate that is available. As shown in ▫ Tables 6.4 and 6.5, this can be achieved after 4 enzymatic steps.

The previous example was relatively simple. We have found a unique possible pathway in the database and the only found branching point, which occurred at reverse step 2, required of two substrates but malonyl-CoA is naturally produced in the chassis. Therefore, we only needed to look for reactions to produce the other reactant (*E*)-cinnamoyl-CoA, as this metabolite is the only one at this step that is not naturally produced in *E. coli*. In a more general case, calculating all possible reactions can become a very laborious procedure and there is no guarantee that we would eventually find a solution. For that reason

---

7 ► https://metacyc.org/compound?orgid=META&id=CPD-6991

Table 6.5 Reactions in the retrosynthetic search for pinocembrin

Step	Reaction
1	Pinocembrin chalcone → (2S)-pinocembrin + H$^+$
2	3 malonyl-CoA + (E)-cinnamoyl-CoA + 3 H$^+$ → pinocembrin chalcone + 3 CO_2 + 4 CoA
3	trans-cinnamate + ATP + CoA → (E)-cinnamoyl-CoA + AMP + diphosphate
4	L-phenylalanine → trans-cinnamate + ammonium

there are several software tools that have been developed to help us in order to determine the reactions that are involved in routes of synthesis. There are **retrosynthesis tools** available from organic chemistry such as `Reaxys`[8] and `Chematica`[9] [3]. Such tools are useful, especially if we would like to consider *semisynthesis* or partial chemical synthesis, i.e. if we use first a biosynthetic pathway in order to produce a natural product as the starting materials to produce other derivative compounds. Here, we would like to focus on biosynthetic pathways involving only enzymatic steps that can be therefore expressed in a chassis organism. For that purpose, **bioretrosynthesis tools** allow pathway screeening and design. One of these tools is `RetroPath` [2]. `RetroPath` uses a representation of the reactions that is similar to the reaction rules that we saw in ▶ Chap. 4, see for instance ◘ Fig. 4.2. In that way, `RetroPath` has the powerful ability of generating putative de novo pathways predicted based on enzyme promiscuity. At the time of this writing, the current version of `RetroPath` is `RetroPath2.0` [1]. It can be easily run within the `KNIME`[10] workflow environment [4]. `RetroPath2.0` can be downloaded from the `MyExperiment` repository.[11] You will need to install these software in order to follow the next calculations in this section. The steps are as follows: (a) install `KNIME` (cross-platform system in `Java`); (b) download the `RetroPath2.0` workflow; (c) Download the `RetroRules` reaction rules.

In order to run `RetroPath`, we need to set up several parameters and define the data sets:

1. **Target compound**: the chemical structure of the target compound to be produced, in our example pinocembrin. The target or targets are defined in a `CSV` file (comma separated values, which can be generated from any spreadsheet) with two columns: "Name" (the desired name for the compound such as "pinocembrin") and "InChI" containing the InChI representation of the molecular structure, as discussed in ▶ Sect. 4.2, which can be easily computed through `RDKit` of `KNIME` starting from the `SMILES` representation of the compound or retrieved from databases such as `ChEBI`;

2. **Sink compounds**: the compounds that are available precursors in the chassis organism. The format is the same as for the target compounds, i.e., a `CSV` file containing two columns: "Name" and "InChI". Genome-scale models, described in ▶ Chap. 2, can be used in order to obtain the list of metabolites that are present in

---

8   ▶ https://www.reaxys.com/
9   ▶ http://chematica.net/
10   ▶ https://www.knime.com
11   ▶ https://www.myexperiment.org/workflows/4987.html

chassis organisms such as *E. coli, S. cerevisiae*, etc. You can obtain the mapping to the chemical structures from `Metanetx`[12] or use `MetaCyc`[13] for that purpose if they are not available in the SBML model of the organism. In the data downloaded from `RetroPath2.0` there is the file `ecoli-iJO1366-mnx-compounds.csv`, which contains the precursors from *E. coli* according to the `iJO1366`[14] model [8] already discussed in ▶ Sect. 2.2;

3. **Reaction rules**: the set of generalized reactions represented as SMARTS or SMIRKS strings (see ▶ Sect. 4.2) at different specificity based on diameter of atomic neighborhood around the reaction center. As discussed in ▶ Sect. 5.3, we can use the `RetroRules` set [2], which is a comprehensive set of biochemical reaction rules. The `RetroRules` set can be downloaded in a format that is ready to be used by the `RetroPath2.0` workflow[15];

4. **Pathway length**: the number of reaction steps in the pathway. The longer we choose the pathway, the longer it would take to compute the solutions. Computation time can grow quickly with each iteration. Typically, any thing beyond 5 steps pathways can easily require more than 24 h of a computation node. At each step, new chemical compounds are generated that need to be tested through retrosynthesis;

5. **Rule diameters**: maximum and minimum rule diameter. Maximum and minimum diameters help us to explore the metabolic space. As we saw in ▶ Chap. 4, a low diameter implies more putative products that are predicted for the substrate based on the reaction rules. Sometimes, we might be interested in exploring such low diameter promiscuous predictions. Lowering the maximum diameter would also help in order to explore highly promiscuous reactions and the production of novel new-to-nature compounds, what is often known as the *underground metabolism* [7];

6. **Number of structures to keep**: `RetroPath` only searches to connect a discrete number of structures at each iteration, which are selected based on the likelihood as promising leads. Because of the highly combinatorial complexity of retrosynthesis at least 100 compounds should be kept, but expanding to 1,000 compounds is a recommended strategy;

7. **Maximum molecular weight**: an often found issue with retrosynthesis algorithms is that they output polymerization results, i.e., the systematic addition of two or more basic monomers or building blocks into a product, which can grow indefinitely based on reaction rules representation. In order to limit this issue, this parameter puts a threshold on the molecular weight of the product generated by the reaction rule.

We are now going to run `RetroPath` for our target pinocembrin by configuring the `KNIME` workflow. As shown in ◳ Fig. 6.1, there are three nodes in the workflow. In the input configuration node we need to fix the parameters of the retrosynthesis described in the previous list. In ◳ Table 6.6 the parameters that were used in the input configuration are listed, where the rest of parameters were left at their default values. In the output configuration node, we need to provide the folder where the results are going to be stored.

After running the program for some time (the calculations can take quite long in cases involving a large number of reaction rules and precursor metabolites and therefore should

---

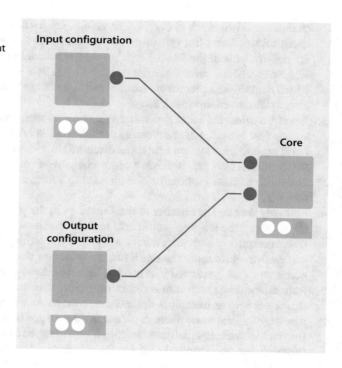

Fig. 6.1 The `Retro-Path2.0` KNIME workflow consists of three nodes: (**a**) input configuration; (**b**) output configuration; and (**c**) core

Input configuration

Core

Output configuration

Table 6.6   Configuration of `RetroPath2.0`

Parameter	Value
Source file	A CSV file containing name and InChI for pinocembrin
Rules file	Reaction rules downloaded from `RetroRules`
Sink file	Metabolites in the `E. coli` iJO1366 model
Pathwhay length (iterations)	6
Minimum rule diameter	8

be preferably based on cluster parallel computing) it will generate a **metabolic scope** file if a solution was found. The scope file can be interactively visualized by using the `scope viewer` tool that comes with `RetroPath2.0`. ◘ Figure 6.2 represents the set of reactions that were found to be involved in some of the pathways connecting pinocembrin to *E. coli* by `RetroPath2.0`. The metabolic scope is the network that connects a target to a source set (see ▶ Box 6.2). The natural pathway given in ◘ Table 6.5 is highlighted on the graph. In total there are 30 reactions in the scope, a number that already starts to become quite complex to analyze in order to identify all pathways contained in the scope. In many cases, the number of reactions in the scope can become very large and therefore it is necessary to apply some algorithm that will look for all possible pathways contained in the scope, as will be discussed in ▶ Chap. 7.

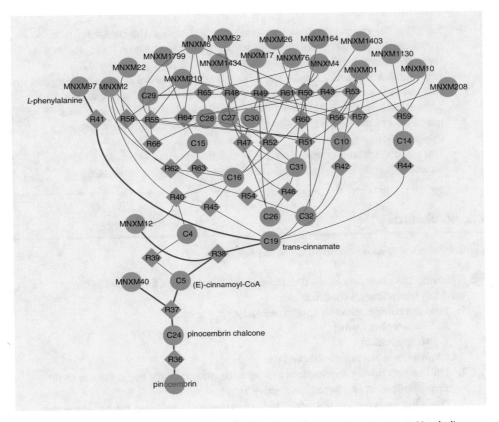

☐ **Fig. 6.2** Metabolic scope of pinocembrin to *E. coli* as generated by `RetroPath2.0`. Metabolites are represented as circles, in green the ones present in the chassis with their corresponding identifiers according to `MetaNetX`; reactions are represented as diamonds. The natural pathway from ☐ Table 6.5 appears highlighted in blue

---

Box 6.2
The **metabolic scope** is a network formed by reactions that has some well-defined properties:
1. Heterologous to the chassis: reactions in the scope are not present in the chassis;
2. Essential to some pathway: reactions in the scope are at least present in one pathway.

`RetroPath2.0` provides also a collection of pre-computed pathways for most of the compounds that are known to have been produced through metabolic engineering. Pinocembrin is among that list of compounds. Therefore, we can also inspect the predictions given by `RetroPath` for the production of pinocembrin in the designated chassis *Escherichia coli* by looking at the pre-computed set. Note that the pre-computed set does not necessarily provide the same solution as in ☐ Fig. 6.2, since this depends on the chosen parameters in ☐ Table 6.6. Often, resulting metabolic scopes can become highly complex. In next ► Chap. 7 we will learn several techniques allowing enumeration and selection of pathways within a retrosynthetic metabolic scope.

┌─ Take Home Message ───┐

— Selecting the chemical target is an important decision at the beginning of a
  metabolic pathway engineering project that needs to consider multiple criteria
  through techno-economic and life cycle analysis.
— Any pathway discovery requires of a careful selection of parameters and desired
  properties in order to narrow down the number of candidate compounds.
— Existence and feasibility of a metabolic scope connecting from the chassis
  precursors to the target chemical through biosynthetic steps are analyzed
  through bioretrosynthesis.
— The use of generalized reaction rules in the bioretrosynthesis analysis allows
  including putative or hypothetical routes based on enzyme promiscuity.

└───┘

6.3 Problems

? 6.1 Select targets based on different requirements.

✓ Following the same procedure described in this section look for targets in ChEBI
verifying the following specifications:
1. Find the compound with applications as
    a. Sweetening agent;
    b. Nutraceutical,
    having the lowest molecular weight;
2. Find targets having applications as fragrance, and cosmetic having at the same
    time application as a hepatoprotective agent;
3. Find all targets having the role of antihypertensive agents and nephroprotective
    agents verifying Linpinski's druglikeness criteria;
4. Select from previous targets those having the highest market price.

? 6.2 Scope at different pathway lengths.

✓ As shown in ◼ Fig. 6.2, the metabolic scope grows as we extend the allowed number
of steps in the pathway. Take some target compound from the previous exercise and
run `RetroPath2.0` for different pathway lengths. Do you see a significant increase in
the size of the scope at higher pathway lengths?

? 6.3 Calculate available metabolic scopes.

✓ In 2004, the National Renewable Energy Laboratory (NREL) from the U.S. Department
of Energy published a report with the top value added chemicals from biomass.[16]
Some of these top compounds are not naturally produced in *E. coli* but are available in
the pre-computed list provided by RetroPath. How many of them can you identify?
Using the provided scope visualizer, create a graphical representation of the metabolic
scopes of the target compounds. Which one has the largest scope?

16 ▶ https://www.nrel.gov/docs/fy04osti/35523.pdf

? 6.4 Screen for other metabolic scopes.

✓ Take the top compounds from the NREL list discussed in the previous question. For some of then, run `RetroPath2.0` at different parameter configurations. Are you able to find additional solutions to the ones in the pre-computed list?

References

1. Delépine, B., Duigou, T., Carbonell, P., Faulon, J.L.: RetroPath2.0: a retrosynthesis workflow for metabolic engineers. Metab. Eng. **45**, 158–170 (2018). https://doi.org/10.1016/j.ymben.2017.12.002
2. Duigou, T., du Lac, M., Carbonell, P., Faulon, J.L.: RetroRules: a database of reaction rules for engineering biology. Nucl. Acids Res. (2018). https://doi.org/10.1093/nar/gky940
3. Feng, F., Lai, L., Pei, J.: Computational chemical synthesis analysis and pathway design. Front. Chem. **6**, 199 (2018). https://doi.org/10.3389/fchem.2018.00199
4. Fillbrunn, A., Dietz, C., Pfeuffer, J., Rahn, R., Landrum, G.A., Berthold, M.R.: KNIME for reproducible cross-domain analysis of life science data. J. Biotechnol. (2017). https://doi.org/10.1016/j.jbiotec.2017.07.028
5. Hastings, J., Owen, G., Dekker, A., Ennis, M., Kale, N., Muthukrishnan, V., Turner, S., Swainston, N., Mendes, P., Steinbeck, C.: ChEBI in 2016: improved services and an expanding collection of metabolites. Nucl. Acids Res. **44**(D1), D1214–D1219 (2016). https://doi.org/10.1093/nar/gkv1031
6. Lipinski, C.A.: Lead- and drug-like compounds: the rule-of-five revolution. Drug Discov. Today Technol. **1**(4), 337–341 (2004). https://doi.org/10.1016/J.DDTEC.2004.11.007; https://www.sciencedirect.com/science/article/pii/S1740674904000551?via%3Dihub
7. Notebaart, R.A., Kintses, B., Feist, A.M., Papp, B.: Underground metabolism: network-level perspective and biotechnological potential. Curr. Opin. Biotechnol. **49**, 108–114 (2017)
8. Orth, J.D., Conrad, T.M., Na, J., Lerman, J.A., Nam, H., Feist, A.M., Palsson, B.O.: A comprehensive genome-scale reconstruction of Escherichia coli metabolism-2011. Mol. Syst. Biol. **7**, 53 (2011). https://doi.org/10.1038/msb.2011.65

Further Reading

A framework for **chemical target** selection:
Campodonico, M.A., Sukumara, S., Feist, A.M., Herrgård, M.J.: Computational methods to assess the production potential of bio-based chemicals. In: Methods in molecular biology (Clifton, NJ), vol. 1671, pp. 97–116 (2018)

A more generic discussion about economic considerations for **bioprocess commercialization**:
Wynn, J.P., Hanchar, R., Kleff, S., Senyk, D., Tiedje, T.: Biobased technology commercialization: the importance of lab to pilot scale-up. In: Metabolic Engineering for Bioprocess Commercialization, pp. 101–119. Springer International Publishing, Cham (2016)

A discussion about **biocatalytic retrosynthesis**:
Green, A.P., Turner, N.J.: Biocatalytic retrosynthesis: redesigning synthetic routes to high-value chemicals. Perspect. Sci. **9**, 42–48 (2016). https://doi.org/10.1016/j.pisc.2016.04.106

A review about **pathway design tools**:
Wang, L., Dash, S., Ng, C.Y., Maranas, C.D.: A review of computational tools for design and reconstruction of metabolic pathways. Synth. Syst. Biol. **2**(4), 243–252 (2017)

Pathway Selection

© Springer Nature Switzerland AG 2019
P. Carbonell, *Metabolic Pathway Design*, Learning Materials in Biosciences,
https://doi.org/10.1007/978-3-030-29865-4_7

What You Will Learn in This Chapter

Several biosynthetic routes are often discovered when screening for ways of connecting the chemical target to the chassis organism through retrosynthesis. The number of alternative pathways can be very high. Pathway enumeration techniques allow determining the alternative solutions in a systematic way. In this chapter, we will introduce elementary flux mode analysis as one of the most successful approaches for pathway enumeration. Those enumerated pathways need to be ranked in order to select the best candidate pathways to engineer. Pathway selection is a multiobjective optimization problem. Different criteria can be used for pathway selection, among others that we will discuss here those main factors due to enzyme performance, production capabilities and intermediate metabolite toxicity.

7.1 Pathway Enumeration

The problem of finding all possible pathways contained within a metabolic scope such as the one shown in ◘ Fig. 6.2 is the problem of **pathway enumeration**. Because its solution can be highly combinatorial, the problem is often constrained by requiring enumerated pathways to be **minimal** (see ▶ Box 7.1). Another important consideration is that we cannot find the pathways by simply traversing the network from the target chemical to the initial precursors like visiting the branches of a tree. The main difference is that metabolic reactions often involve multiple substrates. Therefore, every time that we find one of these reactions with multiple substrates, we need to make sure that each of the substrates are available. Such requirement complicates the calculations. When the problem is approached by using graph theory, enumerating the solutions requires the use of **bipartite graphs or hypergraphs**, complicating the approach beyond the more classical problem of graph theory of finding routes connecting two nodes in a graph.

Box 7.1

A **minimal pathway** is defined as a metabolic pathway where all reactions are essential, i.e., blocking any reaction from the pathway prevents target product. Note that this definition is not necessarily the same as the one that sometimes is used for pathways. A set of reactions considered to form a pathway might be non-minimal. This is for instance the case in many natural pathways that contain some form of redundancy by allowing alternative ways for producing the precursors.

In metabolic engineering we are generally interested in minimal pathways for practical reasons. For instance, introducing additional external enzymes in the cell might create some undesired burden. However, there are cases where some reaction redundancy can be useful. For instance we may want to increase the availability of some natural precursors such as the malonyl-CoA in our pinocembrin example because the amount of precursor that is naturally produced in the cell is not enough for our requirements. In that case in order to increase the final titers, it is licit to introduce an alternative way for producing malonyl-CoA. Even if the resulting pathway might be seen as non-minimal based on the previous definition, from the point of view of pathway performance it should become a preferred solution.

For instance, a simple metabolic scope as the one shown in ◘ Fig. 7.1a can give us an idea of the challenging issues associated with pathway enumeration. In this example, there are multiple reactions requiring several substrates, i.e., R_1, R_2, R_3, R_4 and R_5. Some of them require multiple substrates that are already available in the chassis and therefore do not

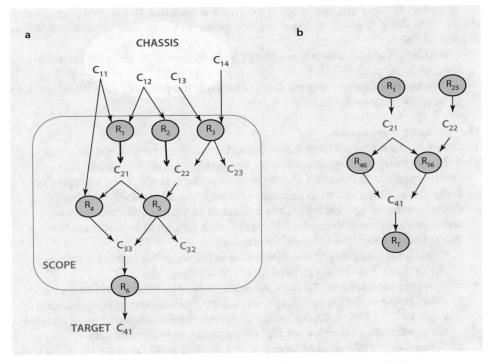

□ **Fig. 7.1** Example metabolic scope for pathway enumeration. **a** Initial metabolic network; **b** Reduced metabolic network for the EFM analysis

increase the problem complexity. However, we can see some cases like R_4 and especially R_5, which requires two intermediates C_{21} and C_{22} that are not naturally produced in the chassis. In such a case, the problem has to be split into two sub-problems, as we need to search for routes that can guarantee the production of each of the two intermediates. Obviously, this will also considerably increase the number of possible pathways because any pathway producing C_{21} could potentially be combined with any pathway producing C_{22}.

Even if the hypergraph approach is a valid way of solving the pathway enumeration, we will consider here another approach to pathway enumeration generally more tractable that is based on metabolic flux analysis. More precisely, we will take advantage of one technique associated with flux balance analysis, which is **elementary flux mode (EFM) analysis**. Each EFM represents a possible different route that steady-state fluxes in the network can take. There are well-known algorithms to compute the EFMs [7]. However, they are computationally expensive and they can take long time to compute and have large memory requirements. On the positive side, they allow parallel computing in cloud computing systems. It is true that most of the complexity occurs when the EFM analysis is performed within the central metabolism because the metabolic network is dense and highly connected. The good news is that here we want to apply EFM analysis to a metabolic scope consisting of heterologous reactions that are imported into the chassis in order to produce some target compound. Therefore, we might expect that the metabolic scope is not so densely connected and we should be able to calculate the pathways through EFM analysis without the level of difficulty seen when dealing with the central metabolism.

The procedure for enumerating the pathways is provided in the following example shown in ■ Fig. 7.1, which will use two computational tools that were introduced in ▶ Chaps. 2 and 3:

- **Construction of the network**: in order to build the metabolic network we will use the `cobrapy` package [3];
- **Computation of elementary modes**: the network analysis tool `COPASI` [4] will help us in order to compute the EFMs.

The procedure is as follows:

1. **Compute the metabolic scope** connecting the target to the chassis either manually or by using some retrosynthetic tool such as `RetroPath`, as discussed in the previous section. ■ Figure 7.1 shows the retrosynthetic scope of our example that connects our target to chassis precursors through several metabolic reactions;
2. **Construct the stoichiometric matrix** equivalent to the metabolic network, where each row corresponds to one metabolite and each column to one reaction. We can distinguish between five classes of metabolites:
   a. **Chassis metabolites (X)**: the metabolites available in the chassis, we assume in this analysis that there is no need for overproducing them;
   b. **Currency metabolites (Y)**: these are readily available metabolites that appear in many reactions, generally acting as co-factors or co-substrates. For instance ATP, NAD, or H_2O. Generally, currency metabolites should be ignored when enumerating pathways, as they are not part of the backbone of the pathway;
   c. **Side products (S)**: metabolites that are a side product of the reactions in the scope but do not play a role in the pathway. They are often currency metabolites as well;
   d. **Pathway intermediates (I)**: heterologous metabolites in the scope that need to be produced in at least one of the producing pathways;
   e. **Target metabolites (T)**: the final product (or products) that we want to produce through the pathways.
3. Remove rows in the stoichiometric matrix representing **chassis compounds**, since we assume that they are already available. In our example, rows C_{11}, C_{12}, C_{13}, C_{14} are removed from the stoichiometric matrix;
4. Remove rows representing **side products** of the reactions, since they are not going to be consumed by any pathway and therefore are irrelevant here. Rows C_{23}, C_{32} will be removed in the example;
5. **Merge reactions** that are identical into a single reaction. Identical reactions can appear once side products, co-factors and chassis metabolites have been removed. These reactions are considered topologically equivalent in terms of pathway enumeration and therefore can be merged into a single column. In our example, columns R_2 and R_3 are merged into column R_{23};
6. **Collapse production of metabolites in cascade reactions** into a single reaction. This step is optional. The rationale for this is that from a topological point of view a linear chain of reactions can be lumped into a single one as long as we are only interested in the end product. In the example, metabolite C_{31} and target metabolite C_{41} can be collapsed by merging reaction R_6 with R_4 and R_5 into reactions R_{46} and R_{56};
7. Add an additional **efflux reaction for the target compound** in order to balance the network so that the target is extracted from the cell.

Equation 7.1 shows the initial stoichiometric matrix for the metabolic network shown in ◘ Fig. 7.1a. Rows and columns that should be merged or collapsed following the described procedure are highlighted in the matrix leading to the network shown in ◘ Fig. 7.1b. Equation 7.2 shows the resulting matrix after reduction of the highlighted cells. An additional efflux reaction for the target reaction R_T was also added to the matrix. In order to calculate the EFMs, we will first build the metabolic model using `cobrapy` as shown in Code 7.1.

Class		R_1	R_2	R_3	R_4	R_5	R_6
X	C_{11}	−1	0	0	−1	0	0
X	C_{12}	−1	−1	0	0	0	0
C	C_{13}	0	0	−1	0	0	0
C	C_{14}	0	0	−1	0	0	0
S	C_{23}	0	0	1	0	0	0
S	C_{32}	0	0	0	0	1	0
I	C_{21}	1	0	0	−1	1	0
I	C_{22}	0	1	1	−0	−1	0
I	C_{31}	0	0	0	1	1	−1
T	C_{41}	0	0	0	0	0	1

$$(7.1)$$

	R_1	R_{23}	R_{46}	R_{56}	R_T
C_{21}	1	0	−1	−1	0
C_{22}	0	1	0	−1	0
C_{41}	0	0	1	1	−1
EFM_1	1	0	1	0	1
EFM_2	1	1	0	1	1

$$(7.2)$$

■■ Code 7.1 Metabolic model of ◘ Fig. 7.1b.

```python
import cobra
model = cobra.Model('Example')
c21 = cobra.Metabolite('C21',compartment = 'c')
c22 = cobra.Metabolite('C22',compartment = 'c')
c41 = cobra.Metabolite('C41',compartment = 'c')
r1 = cobra.Reaction('R1')
r1.add_metabolites({c21: 1.0})
r23 = cobra.Reaction('R23')
r23.add_metabolites({c22: 1.0})
r46 = cobra.Reaction('R46')
r46.add_metabolites({c21: −1.0, c41: 1.0})
r56 = cobra.Reaction('R56')
r56.add_metabolites({c21:−1.0, c22: −1.0, c41: 1.0})
rt = cobra.Reaction('RT')
rt.add_metabolites({c41: −1.0})
model.add_reactions([r1,r23,r46,r56,rt])
model.objective = rt
for r in [r1,r23,r46,rt]:
    r.lower_bound = 0
    r.upper_bound = 1
cobra.io.write_sbml_model(model, 'data/example.xml')
```

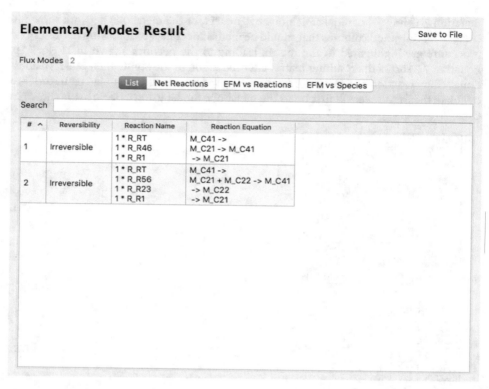

The metabolic network is saved as an SBML model and can then be read by several software tools in order to perform the EFM analysis. Tellurium [2], MetaTool [7], efm-tool [6] are good examples of powerful tools for computing EFMs. For our example in ◻ Fig. 7.1 we will use COPASI in order to perform the EFM analysis. This task can be found under Tasks/Stoichiometric Analysis/Elementary Modes. The result of the EFM analysis is shown in ◻ Fig. 7.2. The system was able to identify two EFMs, which are the ones also shown in Eq. 7.2.

Finally the EFMs calculated in the reduced network of Eq. 7.2 need to be **unfolded into the actual pathways** in the initial scope network of Eq. 7.1:

1. First, **merged reactions** are now unfolded by adding multiple replicates of the folded reaction. In our example, R_{23} is unfolded into R_2 and R_3, which are identical columns. From the point of view of the EFMs, those containing the folded reaction (EFM$_2$ in our case) which contains R_{23}, will generate as many distinct pathways as unfolded reactions, in our case one going through R_2 and the other one through R_3;

2. **Collapsed metabolites** in linear chains are now expanded. In our case, R_{56} is now unfolded into R_5 and R_6 and C_{31} is brought back into the network;

3. Finally, the **target efflux reaction** R_T can be ignored in order to summarize the pathway enumeration.

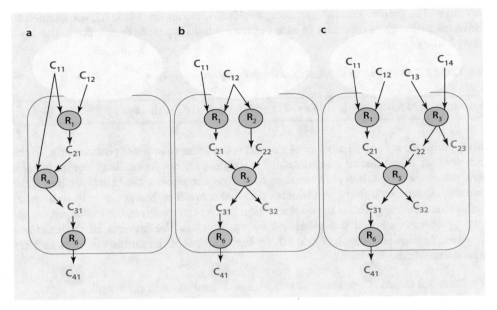

◘ Fig. 7.3 Enumerated pathways for the example network. (**a**) Pathway 1; (**b**) Pathway 2; (**c**) Pathway 3

After applying the previous steps, we finally obtained 3 pathways as shown in Eq. 7.3. The resulting pathways are represented in the network in ◘ Fig. 7.3.

	R_1	R_2	R_3	R_4	R_5	R_6
C_{21}	1	0	0	−1	−1	0
C_{22}	0	1	1	0	−1	0
C_{31}	0	0	0	1	1	−1
C_{41}	0	0	0	0	0	1
P_1	1	0	0	1	0	1
P_2	1	1	0	0	1	1
P_3	1	0	1	0	1	1

$$(7.3)$$

After this description with a simple example of the algorithm that is behind pathway enumeration, return now to our target of pinocembrin production and to its metabolic scope shown in ◘ Fig. 6.2. In order to enumerate the pathways generated by `RetroPath2.0`, we could convert the scope into a SBML file and run any of the described tools such as `COPASI`. Alternatively, `RetroPath2.0` comes with a script based on the `efmtool` that can readily process the output of the `KNIME` workflow. The code `rp2paths` can be downloaded from `GitHub`[1] and will require the installation of some additional Python packages as described in the `GitHub` repository. Note that the following example requires running a Python script, which you should be able to perform either from a shell terminal in your operating system or using a `Jupyter` notebook or the `Spyder` environment, etc.

Once `rp2paths` has been successfully installed, we use the output file from `RetroPath2.0` (`results.csv`) as input in order to perform the enumeration, whose

---

output will be generated in the `pathways` folder. After running Code 7.2, we obtained 60 different pathways (a depiction of each pathway should appear in the output folder as an `svg` folder).

■■ **Code 7.2 Running the `RetroPath` pathway enumeration.**

```
python RP2paths.py all results.csv— outdir pathways
```

Sixty pathways are a large number of alternatives in order to produce pinocembrin. In the next ► Sect. 7.2 we would discuss several strategies that can be applied in order to **rank and prioritize the pathways** in our design. The most straightforward consideration is the number of steps. Intuitively, we should expect that shortest pathways would be more amenable to engineering. This is some information that we can easily read by looking at the file `out_paths.csv` that was generated by `rp2paths`. The file lists all the reactions involved in each pathway. In Code 7.3, we loop and count the number of steps that are associated with each pathway.

■■ **Code 7.3 Count the number of steps for each enumerated pathway.**

```
import csv
pstat = {}
with open( os.path.join('pathways', 'outpaths.csv') ) as handler
    :
    cv = csv.DictReader(handler)
    for row in cv:
        path_id = row['Path ID']
        if path_id not in pstat:
            pstat[path_id] = 1
        else:
            pstat[path_id] += 1
for i in sorted(pstat, key=lambda x: (pstat[x], x)):
    print(i, pstat[i])
```

Interestingly, `RetroPath2.0` found pathways of different length as shown in ▫ Table 7.1. Remember that in ▫ Table 6.6 we limited the number of iterations to 6 and therefore the maximum pathway length is of 7. As proposed, we focus on the pathways involving less enzymatic steps which are the two pathways of length 4 shown in ▫ Fig. 7.4. The first pathway (▫ Fig. 7.4a) is the one that uses *L*-phenylalanine as its precursor, which is converted into cinnamic acid (labeled as `CMPD_50` in the figure) by the PAL enzyme, shown as reaction `MNXR106649` according to the `MetaNetX` reference database. This pathway is the natural route for production of pinocembrin and was the one that we already found in ▫ Table 6.4 by looking at metabolic databases. The second alternative route, shown in ▫ Fig. 7.4b is a route that uses 3-phenylpropanoate as precursor. 3-phenylpropanoate which is actually toxic to *Escherichia coli* can be catabolized by some strains and used as a carbon source. However, this second pathway is not the best candidate as the first pathway uses a proteinogenic amino acid, *L*-phenylalanine that is naturally produced in *Escherichia coli*. Moreover, availability of *L*-phenylalanine can be increased either by overexpressing its biosynthetic pathway or by introducing genetic modifications in the competing path-

□ Table 7.1 Number of pathways found by retrosynthesis in function of the pathway length. In bold the selected pathways of length 4

Pathway length	Number of pathways
4	**2**
5	10
6	24
7	24

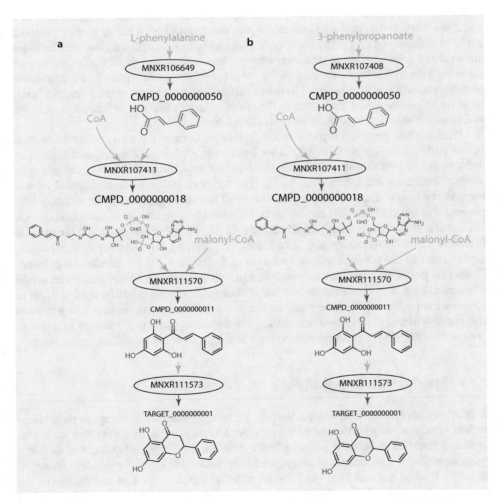

□ Fig. 7.4 Enumerated pathways of length 4 involving two different precursors: **a** *L*-phenylalanine and **b** 3-phenylpropianoate

ways consuming the precursor. We will discuss later in ▶ Chap. 9 different approaches that are possible in order to optimize the chosen pathway from the enumeration. Here in this example, we decided to engineer the route that naturally exists for pinocembrin because of its simplicity and the availability of precursors. In general, would it be possible to find some guidelines in order to select the best pathway(s) from the list of enumerated ones? We would try to address this question in the next section.

7.2 Pathway Ranking

Through this chapter we have defined several steps that can help us in order to systematically search for alternative biosynthesis pathways that can be engineered in chassis organisms to produce a target compound. Finally, an important step for the pathway design is to decide which pathways are more promising or which ones should be prioritized in a metabolic engineering project. Contrary to previous steps where a rather generic protocol could be defined, ranking pathways is a complex task with no single solution because there are so many possible factors that intervene in the decision and there is no easy way of knowing a priori which ones would play a more prominent role. We are facing a problem of **multi-objective optimization**. Therefore, rather than trying to provide a single criteria, what we will do here is to first identify those major factors that can influence the success and to assess how easy is to quantify their effects on the pathway.

   An important point to keep in mind is that we are currently at a **critical decision step** in our pathway design process. So far, we have been applying computational models in order to get an overall vision of the alternatives that exist for implementing the pathway. We have had not started yet to truly design the pathway, which is something that we will learn in the next ▶ Chap. 8. At this point we need to make the assumption that choosing the pathway is a decision that can be made somehow independently of the actual design, i.e., the actual selection of the genetic parts in the pathway. Pathway selection determines therefore a first decision step and once it has been decided, any pathway design coming downstream in the pipeline will be based on such decision and on the chosen pathway. From this point, downstream pathway design will be devoted to designing the actual implementation of the pathway and several optimization strategies will be used in order to improve the performance of the pathway. But first what we want to do here is **to estimate the capabilities of the pathways** and their theoretical optimal performance in order to prospectively select the most promising ones based on their expected theoretical limits.

7.2.1 Production Capabilities

A relatively straightforward comparison for pathways would be to compare their **theoretical maximum production and biomass**. For instance, in the examples that we saw in ▶ Sect. 7.1, each enumerated pathway had a different topology and consumed different precursors. For the pinocembrin target, we finally decided to select the pathway shown in ▣ Tables 6.4 and 6.5 and in ▣ Fig. 7.4a. In order to simulate the flux capabilities of the chassis for this pathway we will define the metabolites and reactions of the pathway using cobrapy in Code 7.4 in the same way as was done for the example in ▶ Sect. 7.1. As we did in the previous example, we will need to add an additional efflux reaction R_T for the target.

■■ **Code 7.4 Definition of the pinocembrin metabolic pathway.**

```
import cobra
pino = cobra.Metabolite('pino_c', compartment='c')
pinocha = cobra.Metabolite('chal_c', compartment='c')
maloncoa = model.metabolites.get_by_id( "malcoa_c")
cinnacoa = cobra.Metabolite('cinn_c', compartment='c')
coa = model.metabolites.get_by_id "coa_c")
cinn = model.metabolites.get_by_id ( "cinnm_c")
co2 = model.metabolites.get_by_id ( "co2_c")
atp = model.metabolites.get_by_id ( "atp_c")
amp = model.metabolites.get_by_id ( "amp_c")
ppi = model.metabolites.get_by_id ( "ppi_c")
phen = model.metabolites.get_by_id( "phe__L_c")
nh4 = model.metabolites.get_by_id ( "nh4_c")
pal = cobra.Reaction('PAL ')
pal.add_metabolites({phen: -1.0, cinn: 1.0, nh4: 1.0})
r4cl = cobra.Reaction('4CL ')
r4cl.add_metabolites({cinn: -1.0, atp: -1.0, coa: -1.0, cinnacoa:
     1.0, amp: 1.0, ppi: 1.0})
chs = cobra.Reaction('CHS ')
chs.add_metabolites({maloncoa: -3.0, cinnacoa: -1.0, pinocha:
     1.0, co2: 3.0, coa: 4.0})
chi = cobra.Reaction('CHI ')
chi.add_metabolites({pinocha: -1.0, pino: 1.0})
rt = cobra.Reaction('RT ')
rt.add_metabolites({pino: -1.0 })
for r in [pal, r4cl, chs, chi, rt]:
    r.lower_bound = 0
    r.upper_bound = 1000
```

The pathway will be then imported into the chassis *E. coli* by adding the pathway reactions to the model as shown in Code 7.5.

■■ **Code 7.5 Definition of the metabolic model of the *E. coli* engineered strain with the pinocembrin pathway.**

```
import cobra.test
model = cobra.test.createtest_model("ecoli")
model.add_reactions([pal, r4cl, chs, chi, rt])
```

The *E. coli* model in `cobrapy` has initially as objective the biomass reaction, as discussed in ▶ Sect. 2.3. In Code 7.6, flux balance analysis gave an optimal solution of `max_bio-mass` = 0.9824 h^{-1}.

■■ **Code 7.6 Calculation of the maximum achievable biomass.**

```
solution = model.optimize()
max_biomass = solution.objective_value
```

We want now to change the objective to our target reaction R_T. Before doing this, we store in Code 7.7 the current biomass objective of the model.

■■ Code 7.7 Calculation of the maximum achievable target flux.

```
from cobra.util.solver import linear_reaction_coefficients
biomassObjective = linear_reaction_coefficients(model)
model.objective = 'RT'
solution = model.optimize()
max_product = solution.objective_value
```

The maximum production for pinocembrin is 3.04 mmol × gDW^{-1} × h^{-1}. We can therefore assume that production of pinocembrin will lie at some value between $[0, 3.04]$ depending on how much flux we are able to divert from other cellular functions into our target production. In order to get an idea of the maximum target production depending on the actual biomass, we vary the target from 0 to its theoretical maximum and optimize for maximum biomass, as shown in ◘ Fig. 7.5.

If our objective is to simultaneously maximize biomass and production, a first estimate will be to look for the theoretical maximum achievable combination of production and biomass in ◘ Fig. 7.5 that maximizes the objective function:

$$J = v_{biomass} \times v_{target} \tag{7.4}$$

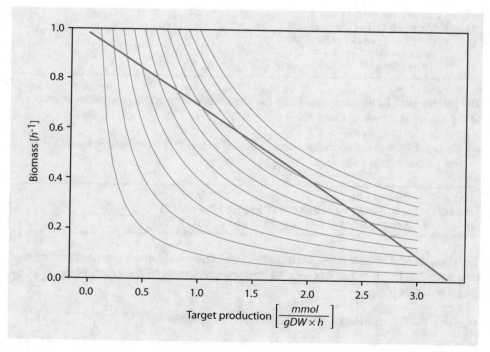

◘ **Fig. 7.5** Comparison between theoretical production of the target and maximum biomass. Curves represent increasing values of the coupled objective J

as shown through a set of isocost curves generated in Code 7.8 each having a constant J. The point that is tangent to the highest isocost curve will be the maximum achievable coupled biomass and production point.

■■ **Code 7.8 Plot optimal curve points.**

```
model.objective = lc
sols = []
n = 100
for x in range(0, n):
    val = x*max_product/n
    rt.lower_bound = val
    rt.upper_bound = val
    solution = model.optimize()
    biomass = solution.objective_value
    sols.append( [val, biomass] )
import matplotlib.pyplot as plt
import numpy as np
sols = np.array( sols )
plt.scatter(sols[:,0],sols[:,1])
plt.xlabel(r'Target production $\left[ \dfrac{mmol}{gDW \times h}
    \right]$')
plt.ylabel(r'Biomass $\left[ h^{-1} \right]$')
plt.show()
```

7.2.2 Enzyme Performance

Besides steady-state production capabilities, **enzyme performance** is another key element in the pathway. Depending on the actual kinetics of the pathway enzymes, equilibrium values can be far from the optimal steady-state values. As was discussed in ► Chap. 5, this is a non-trivial problem and cannot be always estimated in advance. The approach here should be to select best sequence candidates based on the available information in databases and on the enzyme selection tools described in ► Chap. 5. Some generic rules are:

- Try to **avoid bottlenecks** by keeping kinetic constants at the same level;
- **Diversify the selection**, instead of selecting a single enzyme sequence for each biochemical conversion, select a short list of candidate sequences to test following the procedures outline in ► Chap. 5;
- In some cases, it is possible to **select a modular pathway** where different parts are carried out in different compartments of the cell or even in different cells. In that case, the selection of the enzymes should focus on each module rather than on the full pathway.

It is important to assign a score to each of the selected enzymes. The score can be based on the same approach as the one described for `Selenzyme` in ► Chap. 5 or simply on sequence similarity. Some rules in order to generate a sequence score are as follows:

- Sequences that are annotated receive a higher score than sequences that were selected by homology;
- Sequences with lower K_M (affinity to the substrate) and higher k_{cat} (turnover rate) receive a higher score;

— Sequences whose source is phylogenetically closer to the host organism receive a higher score;

— Sequences that are highly conserved within the enzyme family receive a higher score.

7.2.3 Toxicity

Another aspect to consider is the toxicity or inhibition effects of the pathway. A new pathway engineered in the chassis implies heterologous enzymes that are expressed in the cell and heterologous metabolites. We need to make sure that the toxicity of the heterologous enzymes can be alleviated and that intracellular accumulation of new intermediates and target metabolite are kept below concentrations causing significant inhibition effects.

Several databases and tools exist to assess toxicity of chemicals and enzymes in chassis organisms. For instance, EcoliTox[2] [5] is a predictive online tool based on a database of toxicity data in *E. coli*. This toxicity predictive model was developed based on chemical descriptors and could be adapted to other chassis organisms.

> **Take Home Message**
>
> — Pathway enumeration allows determining all possible routes in the metabolic scope connecting the chassis to the target.
> — Elementary flux mode analysis is one algorithmic approach to pathway enumeration.
> — Enumerated pathways need to be ranked in order to identify best candidate pathways to engineer.
> — Enzyme performance, production capabilities and intermediate toxicity are among the criteria used to rank engineered metabolic pathways.

7.3 Problems

? **7.1** Calculate the optimal couple production and biomass point.

✓ Obtain for the pinocembrin model example the optimal solution according to objective *J* in Eq. 7.4.

? **7.2** Calculate yield of target product on substrate.

✓ Assuming that glucose glc_c is our substrate, calculate the yield of pinocembrin (g product per g substrate) at maximum and optimal production according to previous exercise.

? **7.3** Select candidate sequences for the pathway.

✓ Select at least 4 candidate sequences for each step of the pathway in ▢ Table 6.4 and rank each step based on the criteria described in this section. *Hint:* you can use the Selenzyme tool [1] in order to perform the selection.

---

2 ▶ https://absynth.issb.genopole.fr/Bioinformatics/

7.4 Estimate toxicity effects of the pathway.

Using `EcoliTox`, compute the estimated toxicity of each heterologous compound, i.e., for all metabolites that are not present in the *E. coli* model. *Hint:* Select at least 4 candidate sequences for each step of the pathway in ◻ Table 6.4 and rank each step based on the criteria described in this section. *Hint:* the pathway definition using `cobrapy` shown in ▶ Sect. 7.2 provides the list of heterologous metabolites.

References

1. Carbonell, P., Wong, J., Swainston, N., Takano, E., Turner, N.J., Scrutton, N.S., Kell, D.B., Breitling, R., Faulon, J.L.: Selenzyme: enzyme selection tool for pathway design. Bioinformatics **34**(12), 2153–2154 (2018). https://doi.org/10.1093/bioinformatics/bty065
2. Choi, K., Medley, J.K., König, M., Stocking, K., Smith, L., Gu, S., Sauro, H.M.: Tellurium: an extensible python-based modeling environment for systems and synthetic biology. Biosystems **171**, 74–79 (2018). https://doi.org/10.1016/j.biosystems.2018.07.006
3. Ebrahim, A., Lerman, J.A.J., Palsson, B.O., Hyduke, D.R.: COBRApy: COnstraints-based reconstruction and analysis for python. BMC Syst. Biol. **7**(1), 74 (2013). https://doi.org/10.1186/1752-0509-7-74
4. Hoops, S., Sahle, S., Gauges, R., Lee, C., Pahle, J., Simus, N., Singhal, M., Xu, L., Mendes, P., Kummer, U.: COPASIa COmplex PAthway SImulator. Bioinformatics **22**(24), 3067–3074 (2006). https://doi.org/10.1093/bioinformatics/btl485
5. Planson, A.G.A.G., Carbonell, P., Paillard, E., Pollet, N., Faulon, J.L.: Compound toxicity screening and structure-activity relationship modeling in Escherichia coli. Biotechnol. Bioeng. **109**(3), 846–850 (2012). https://doi.org/10.1002/bit.24356
6. Terzer, M., Stelling, J.: Large-scale computation of elementary flux modes with bit pattern trees. Bioinformatics **24**(19), 2229–2235 (2008). https://doi.org/10.1093/bioinformatics/btn401
7. Zanghellini, J., Ruckerbauer, D.E., Hanscho, M., Jungreuthmayer, C.: Elementary flux modes in a nutshell: properties, calculation and applications. Biotechnology J. **8**(9), 1009–1016 (2013). https://doi.org/10.1002/biot.201200269

Further Reading

Hypergraph techniques for metabolic **pathway enumeration**:
Carbonell, P., Fichera, D., Pandit, S., Faulon, J.L.: Enumerating metabolic pathways for the production of heterologous target chemicals in chassis organisms. BMC Syst. Biol. **6**(1), 10 (2012). https://doi.org/10.1186/1752-0509-6-10
Liu, F., Vilaa, P., Rocha, I., Rocha, M.: Development and application of efficient pathway enumeration algorithms for metabolic engineering applications. Comput. Methods Prog. Biomed. (2014). https://doi.org/10.1016/j.cmpb.2014.11.010
An early discussion about **pathway ranking**:
Carbonell, P., Planson, A.G., Fichera, D., Faulon, J.L.: A retrosynthetic biology approach to metabolic pathway design for therapeutic production. BMC Syst. Biol. **5**(1), 122 (2011). https://doi.org/10.1186/1752-0509-5-122
A proof-of-concept of the application of different approaches to pathway selection and ranking appeared in the report of a recent **pressure test** for synthetic biology biofoundries:
Casini, A., Chang, F.Y., Eluere, R., King, A.M., Young, E.M., Dudley, Q.M., Karim, A., Pratt, K., Bristol, C., Forget, A., Ghodasara, A., Warden-Rothman, R., Gan, R., Cristofaro, A., Borujeni, A.E., Ryu, M.H., Li, J., Kwon, Y.C., Wang, H., Tatsis, E., Rodriguez-Lopez, C., O'Connor, S., Medema, M.H., Fischbach, M.A., Jewett, M.C., Voigt, C., Gordon, D.B.: A pressure test to make 10 molecules in 90 days: external evaluation of methods to engineer biology. J. Am. Chem. Soc. **140**(12), 4302–4316 (2018). https://doi.org/10.1021/jacs.7b13292

Metabolic Pathway Design

Contents

Pathway Design

© Springer Nature Switzerland AG 2019
P. Carbonell, *Metabolic Pathway Design*, Learning Materials in Biosciences,
https://doi.org/10.1007/978-3-030-29865-4_8

What You Will Learn in This Chapter

Previous chapters have focused on metabolic pathway, and their genetic parts, modeling and selection. In this chapter, we direct our interest towards pathway design in order to move into the engineering aspects of metabolic pathway design. Our focus from now on will not be on reproducing through simulation some biological behavior or trying to understand its mechanism but in determining the possible biological interventions and modifications to be engineered in the cell in order to achieve the desired behavior. Genetic parts selection and experimental design will be essential at this stage. Since this book is about metabolic pathway design, metabolism will continue as our central topic. However, we will also consider solutions that go beyond metabolism and will take advantage of the advanced capabilities that synthetic biology provides to metabolic engineering.

8.1   Selecting the Functional Genes in the Pathway

We start this chapter at the exact point where we left our previous ▶ Chap. 7:
1. Heterologous enzymatic reactions leading to the target chemical from the chassis have been already identified through retrosynthesis (▶ Chap. 6);
2. Pathways contained in the resulting scope have been enumerated;
3. Pathways have been ranked following the described criteria in ▶ Chap. 7;
4. Target pathway or pathways have been selected based on the ranking.

In the following, we assume that the preceding steps led us to select the pathway shown in ◻ Table 6.4 and ◻ Fig. 7.4a. This pathway involves 4 enzymatic steps to produce the target flavonoid pinocembrin. For each step, we need to **select the candidates enzyme sequences**. Choosing multiple candidates among the top ranked sequences instead of just a top candidate is a recommended strategy as it will increase the chances of finding high producer pathways. In our example, we will consider 4 candidate genes per enzyme, although this is a number that might change depending on each case. In order to help us on the selection of the sequences, we will use the tool `Selenzyme`[1] for enzyme selection, already introduced in ▶ Sect. 5.2. `Selenzyme` will search for annotated sequences for the target reaction and for other close reactions in terms of chemical similarity.

`Selenzyme` provides multiple columns that can be used in order to define a selection score by assigning weights w_i to the ranking score of each column c_i (see in ◻ Fig. 8.1 an example of a typical output for an enzyme query):

$$J_{seq} = \sum w_i c_i \tag{8.1}$$

In our case, the selection will be based on the following criteria listed in descending order of priority:
1. **Reaction similarity:** which is a value in the range of $[0,1]$, will be given the highest priority, with a weight of $w_{rsim} = 10^3$;
2. **Sequence conservation:** in a range of $[0,100]$ will help in order to focus on highly conserved members of the enzyme family. The column weight will be $w_{cons} = 10$;

---

1   ▶ http://selenzyme.synbiochem.co.uk

Results

[Delete selected rows] [Download CSV] [Download FASTA] [Download MSA] [View MSA] [Add custom sequences (FASTA)] [Column definition/Change score]

	Score	Seq. ID	Description	Organism Source	Tax. distance	Rxn. ID	EC Number	Uniprot protein evidence	Consv. Score	Rxn Sim.	Direction Used	Direction Preferred	% helices	% sheets	% turns	% coils	Mol. Weight	Isoelec. Point	Polar %
1	160.8	P45727	Phenylalanine ammonia-lyase	Persea americana	24	MNXR7145	4.3.1.24;4.3.1.25	2	85.0	1.0	-1	-1	43.2	18.7	19.0	21.7	67875.71	6.6014	45.968
2	158.8	P45726	Phenylalanine ammonia-lyase	Camellia sinensis	27	MNXR7145	4.3.1.24;4.3.1.25	2	86.0	1.0	-1	-1	46.0	18.5	17.2	20.6	77751.87	6.5626	45.658
3	156.8	P25872	Phenylalanine ammonia-lyase	Nicotiana tabacum	30	MNXR7145	4.3.1.24;4.3.1.25	2	87.0	1.0	-1	-1	46.8	17.0	15.3	23.2	77780.81	6.7542	45.734
4	156.8	O23924	Phenylalanine ammonia-lyase	Digitalis lanata PE=2 SV=1	29	MNXR7145	4.3.1.24;4.3.1.25	2	86.0	1.0	-1	-1	47.9	18.2	14.9	21.2	77732.73	6.6425	46.003
5	156.7	Q40910	Phenylalanine ammonia-lyase G4 (Fragment)	Populus kitakamiensis	29	MNXR7145	4.3.1.24;4.3.1.25	3	86.0	1.0	-1	-1	44.3	20.5	16.0	22.0	62483.71	6.2428	44.483
6	155.9	P45724	Phenylalanine ammonia-lyase 2	Arabidopsis thaliana	29	MNXR7145	4.3.1.24;4.3.1.25	1	85.0	1.0	-1	-1	42.9	22.1	14.7	22.5	77859.93	6.4177	46.444
7	155.9	P35510	Phenylalanine ammonia-lyase 1	Arabidopsis thaliana	29	MNXR7145	4.3.1.24;4.3.1.25	1	85.0	1.0	-1	-1	43.2	19.6	15.7	23.8	78725.73	6.2471	46.345
8	155.8	Q42667	Phenylalanine ammonia-lyase	Citrus limon	29	MNXR7145	4.3.1.24;4.3.1.25	2	85.0	1.0	-1	-1	41.8	19.1	16.6	24.8	78490.41	6.5555	47.230
9	155.8	O49836	Phenylalanine ammonia-lyase 2	Lithospermum erythrorhizon PE=2 SV=1	30	MNXR7145	4.3.1.24;4.3.1.25	2	86.0	1.0	-1	-1	50.1	16.5	13.1	22.6	76509.51	6.1855	44.681
10	155.8	Q01861	Phenylalanine ammonia-lyase 1	Pisum sativum	30	MNXR7145	4.3.1.24;4.3.1.25	2	86.0	1.0	-1	-1	44.4	20.9	16.7	20.2	78830.79	6.7633	47.026

Fig. 8.1 Output showing the top 10 sequence candidates selected by `Selenzyme` for PAL (EC 4.3.1.24)

3. **Taxonomic distance**: a value that counts the number of nodes between the target host (*E. coli*) and the sequence source in the tree of the NCBI taxonomy,[2] generally a value in the range of $[1,50]$. The associated weight is $w_{tax} = -10$. Note that the score is here negative in order to favor those sequences that originate from a source that is phylogenetically closer to the chassis;
4. **Uniprot protein evidence**: a value between $[1,5]$, with 1 meaning experimental evidence at protein level, 2 at transcript level, 3 inferred by homology. The weight will be again negative in order to give priority to proteins with experimental evidence, with an associated weight of $w_{uni} = -10$;
5. **Protein molecular weight (MW)**: will be also considered in order to avoid fragment sequences as much as possible. The chosen weight is of $w_{mw} = 1$.

Based on the previous criteria, Table 8.1 shows the 4 selected candidate sequences for PAL. The columns correspond to the described values for sequence conservation (*Conservation*), taxonomic distance (*Distance*), Uniprot protein evidence (*Evidence*) and molecular weight (*MW*). All of them came from plant sources and are annotated for the phenylalanine ammonia lyase (EC 4.3.1.24) and therefore they have a reaction similarity of 1.0.

We proceed next with the selection for 4-Coumarate-CoA ligase (4CL) as shown in Table 8.2. In this case we have removed some candidates because those sequences were annotated as "probable" or were homologs in the same species. We should prioritize enzyme sequences that have been experimentally verified for the target activity and also try to increase sequence diversity by not choosing too close homologs.

Similarly, Tables 8.3 and 8.4 show the candidate sequences for chalcone synthase (CHS) and chalcone isomerase (CHI), respectively.

2 ► https://www.ncbi.nlm.nih.gov/taxonomy

Table 8.1   Selected candidate sequences for PAL

Score	Uniprot	Source	EC number	Conservation	Distance	Evidence	MW
1590	P45727	Persea americana	4.3.1.24	85	24	2	67875.71
1560	P45726	Camellia sinensis	4.3.1.24	85	27	2	77751.87
1560	P45735	Vitis vinifera	4.3.1.24	86	28	2	46015.38
1540	P35510	Arabidopsis thaliana	4.3.1.24	84	29	1	78725.73

Table 8.2   Selected candidate sequences for 4CL

Score	Uniprot	Source	EC number	Conservation	Distance	Evidence	MW
1354	P41636	Pinus taeda	6.2.1.12	63	24	2	58590.86
1326	O24540	Vanilla planifolia	6.2.1.12	64	27	3	60095.61
1316	O24145	Nicotiana tabacum	6.2.1.12	65	30	2	59842.33
1314	P31686	Glycine max	6.2.1.12	66	31	2	32027.23

Table 8.3   Selected candidate sequences for CHS

Score	Uniprot	Source	EC number	Conservation	Distance	Evidence	MW
1585	P48387	Camellia sinensis	2.3.1.74	92	27	2	42595.28
1573	Q8RVK9	Cannabis sativa	2.3.1.74	91	28	1	42720.37
1572	P48388	Camellia sinensis	2.3.1.74	91	27	2	42814.67
1562	P51090	Vitis vinifera	2.3.1.74	91	28	2	42701.36

■ **Table 8.4** Selected candidate sequences for CHI

Score	Uniprot	Source	EC number	Conser-vation	Dis-tance	Evi-dence	MW
1648	P28012	Medicago sativa	5.5.1.6	70	30	1	23826.29
1482	P41088	Arabidopsis thaliana	5.5.1.6	82	29	1	26595.61
1414	Q93XE6	Glycine max	5.5.1.6	75	31	2	23264.81
1385	P11650	Petunia hybrida	5.5.1.6	72	29	2	26088.83

8.2 Transcriptional and Translational Tuning

Independently of the assembly method for the DNA parts in the pathway, the next design decision is about the selection of the genetic regulatory elements. When designing a pathway that is going to be expressed in a plasmid introduced into a bacterial host such as *E. coli*, we have the possibility of adding genetic elements that will allow us to tune the response of the pathway by introducing transcriptional and translational regulation.

Translational rates can be modified through alteration of the **ribosome binding site (RBS)** strength. RBSs are regulatory RNA sequences found upstream of the start codon of a coding region. They are often present in bacteria and their role is the recruitment of a ribosome during the initiation of protein translation. Similarly to a natural RBS, a synthetic RBS can be designed in order to control the levels of gene expression. To that end, several predictive models have been developed that relate the RNA sequence to the RBS strength. For instance, the RBS Calculator [6] is generally used in order to design an RBS with a designated strength. In our pathway, it would be desirable to introduce multiple RBSs with different strengths as a way of tuning the expression of the pathway genes.

Transcriptional tuning is another regulatory element that exists in the pathway. **Promoters** are DNA regions that initiate transcription. Multiple promoters have been characterized in different chassis organisms. Some promoters are constitutive, i.e., they are always active. Interestingly, there exist also promoters that are inducible, meaning that the presence of some chemical inducer introduces a change in its activation (see ► Box 8.1). Similarly to translational tuning, it would be desirable to select several promoters with different transcriptional strengths for transcriptional tuning.

For our pathway example, we are going to make the following design choices:

1. **Translational tuning**: RBSs will be individually selected for each gene sequence in order to have a uniform estimated translational rate through the online Ribosome Binding Site Calculator[3];
2. **Transcriptional tuning**: two inducible promoters will be selected as possible design choices, i.e., the lactose-inducible *Ptrc* promoter and the arabinose-inducible promoter *PBAD*.

---

3 ► https://salislab.net/software/

> **Box 8.1**
> The *lac* **promoter** is an inducible promoter found in the *lac* operon, which is an operon involved in the transport and metabolism of lactose in *E. coli*. The *lac* promoter is negative inducible, meaning that normally the promoter is repressed or turned off but in the presence of lactose it is activated.
>
> In bioengineering applications, lactose is replaced by some other analog inducer such as Isopropyl β-D-1-thiogalactopyranoside (IPTG). One advantage of IPTG is that it cannot be metabolized by *E. coli* and therefore its concentration remains constant during induction.

8.3 Combinatorial Design

We have now selected a collection of genes, promoters and the RBSs as part of our design. Besides the gene coding region and the transcriptional and translational regulation, there are several other parameters that can be tuned in the design. ◻ Table 8.5 lists a some of the most common pathway design tuning parameters that can generally be modulated in a pathway design in order to obtain a different response. Most of them are discussed in this chapter or elsewhere in this textbook.

For instance, some of the elements in ◻ Table 8.5 that are generally used in the design space of the plasmid construct are:

- **Origin of replication (ori)**, which is a DNA sequence that directs the initiation of plasmid replication in bacteria by recruiting the transcriptional machinery. The ori allows controlling the **DNA copy number**, i.e., the number of plasmids that are copied (amplified) in a cell. For example, the widely used pBR322 cloning vector produces 30–40 plasmids per cell, while pSC101 produces around 5 copies per cell, or p15A, which produces around 10–20 copies per cell. Higher copy number will lead to higher levels of expression of the enzymes in the pathway;
- **Resistance cassette**, i.e., the antibiotic resistance gene (ampicillin, gentamycin, kanamycin, tetracycline, chloroamphenicol, etc.) that allows for selection of the plasmid-containing bacterial host. Plasmids introduce a replication burden that makes that bacteria without plasmid quickly outgrow the plasmid-containing hosts in bacterial populations. Adding a resistance gene allows ensuring the retention of the plasmid DNA in the bacterial populations.

Similarly, other factors that we might consider in our prototype design include:

- **Chassis**: selection of different chassis can have a radical effect on the choice of design parameters. However, within a selected chassis there are often multiple strains as in the case of *E. coli* where many lab and industrial strains have been developed, each with some specific properties such as those labeled as MG1655 (K12), BL21 (high protein expression), DH5α or DH10β;
- **Growth medium**: growth media can have an important effect on the resulting titers. Some growth medium such as rich media can lead to higher production levels but might also present low reproducibility, while minimal media has good reproducibility but production levels are low. Widely-used growth media for *E. coli* are LB (Lysogeny Broth or Luria-Bertani), TB (Terrific Broth), and M9 (minimal medium).

□ Table 8.5 Selection of pathway design tuning parameters

Level	Parameter
Transcription	Promoter type
Transcription	Inducer concentration
Transcription	Promoter leakiness and basal expression
Transcription	Promoter strength
Transcription	Placement of genes
Translational	Ribosome-binding site (RBS) strength
Translational	Codon optimization
Translational	mRNA decay rate
Translational	Riboregulators
Post-translational	Inteins (protein splicing)
Post-translational	Localization and co-localization
Protein	Protein degradation
Protein	Enzyme activity
Protein	Dynamic balancing
Plasmid	DNA copy number
Plasmid	Resistance cassette
Process	Chassis
Process	Media
Process	Temperature
Process	Induction point
Process	Inducer concentration
Process	Substrate delay
Process	Coculture and polyculture

8.4 Experimental Design

As discussed in previous ▶ Sect. 8.3, the number of possible factors that can be tuned in a pathway design is overwhelming. Unfortunately, it is not always obvious which parameters are going to have a major positive impact in the response. For instance, □ Table 8.6 shows a typical selection of design parameters for our pathway example that would need to be tested. Concerning the promoter selection, it involves 2 possible promoters (*Ptrc* and *PBAD*) at the initial position in front of the PAL gene, while before the second and third

□ Table 8.6   Selection of pathway design tuning parameters

Number	Factor	Levels	Labels
0	resistance	2	ampicillin, kanamycin
1	ori	2	pSC101, pBR322
2	promoter	2	Ptrc, pBAD
3	PAL	4	P45727, P45726, P45735, P35510
4	promoter	3	no, Ptrc, pBAD
5	4CL	4	P41636, O24540, O24145, P31686
6	promoter	3	no, Ptrc, pBAD
7	CHS	4	P48387, Q8RVK9, P48388, P51090
8	CHI	4	P28012, P41088, Q93XE6, P11650
9	Chassis	3	MG1655, DH5α, DH10β
10	Media	3	LB, TB, M9

genes, 4CL, CHS respectively, the possibilities are either no promoter, *Ptrc* or *PBAD*. Finally, there is no promoter in front of the fourth gene (CHI). With such design space, the total number of possible combinations is:

$$2 \times 2 \times 2 \times 4 \times 3 \times 4 \times 3 \times 4 \times 4 \times 3 \times 3 = 52,596$$

(8.2)

a number that most likely will exceed the capabilities and resources of any synthetic biology lab. Therefore, we need to find an approach that helps us on reducing the number of different constructs to build or experiments to run in the lab. Fortunately, **statistical design of experiments** provides techniques that we can use here in order to optimize our design (see ► Box 8.2).

Statistical design of experiments (DoE) tries to overcome the limitations of traditional experiments where only one factor is changed at a time. For instance, in our example with a design space defined by □ Table 8.6, we could first start by trying the two resistance cassettes and keep the one that has better performance, then repeat the procedure for the two origins of replications, followed by the 2 promoters, the 4 PAL genes, etc. However, this approach is not efficient because it is time and resource consuming and, more importantly, because main interactions between factors can be missed. The DoE approach consists on simultaneously screening for a set of combinations of the factors providing an optimal coverage of the design space and therefore allowing the identification of solutions with few experimental runs.

□ Table 8.7 provides an example of an optimal combinatorial library of size 24 designed for screening the design space of □ Table 8.6 in the lab. The library was generated using an optimal design approach. Optimal designs are based on algorithms that look for a possible experimental design solution that minimizes some given criteria. One of the most widely used criteria is to minimize the estimation error for the linear regression parameters of an assumed model between the factors and the response. □ Table 8.6 is obviously just one of

Box 8.2

Statistical design of experiments (DoE) has a rich history from early days in agricultural planning to optimization of industrial manufacturing and chemical processes starting from the 1950s/1960s. From the 1970s/1980s, its application to automotive and electronics industry became widespread. The most basic approach consists of:

1. Define project goal and response variables. In our example response variables are production titers and biomass, initial pilot tests should be performed in order to scope the response variables, i.e. to find their range values;
2. Define factors for experimentation. We need to make distinction between:
   a. **Categorical factors**, such as those that are used in our example in ■ Table 8.6 with multiple labels;
   b. **Numerical factors** like temperature, pH, etc., (see for instance ■ Table 8.5) in that case we need to define the factor levels, generally consisting of a low level, a high level and a center point;
3. Identify factors that we cannot change or control, for instance, room temperature, in that case either we keep them constant or we randomize them;
4. There are many designs available for screening: full factorial, fractional-factorial, Plackett-Burman, optimal, etc.[4] Optimal designs are particularly well-suited for pathway design. They aim at selecting a desired number of runs or experiments maximizing the mutual information content of the design and/or minimizing the estimation error in the assumed model;
5. Commercial DoE software like `JMP` (▶ https://www.jmp.com), `Minitab` (▶ http://www.minitab.com), `Design Expert` (▶ https://www.statease.com/) and others are often used in industrial applications in order to generate a design like the one in ■ Table 8.7. Community-based open-source solutions based on Python and R are increasingly becoming available.

the many possible designs of optimal combinatorial libraries of size 24 that could be generated. Testing a library like the one given in the example of ■ Table 8.6 that is based on statistical principles rather than a randomly (or heuristically) generated library from ■ Table 8.5 should be the preferred strategy because this can save a lot of time and resources by targeting an optimal coverage of the design space.

In general, DoE will be applied for those factors that we cannot change and screen in a high-throughput fashion. For instance, if testing the strain in different media is something that can be automated by using a robotic platform, microfluidics, etc., we should evaluate whether it would be preferable screening all possible combinations. Similarly, DoE provides techniques of **block design**, which allow us to split the experiment into a batch of different runs where in each case one of the factors is kept constant.

Going back into the example of testing the strains in different media, a block design for media will involve designing separate combinatorial libraries to be grown in LB, M9 and TB. Deciding which strategy better works depend on the capabilities of the build platform. When defining the different factors, we need to evaluate their experimental constraints in order to decide if the factor requires of some specific design like:

— If changing the factor can be automated, we might opt for testing all possible combinations of the factor: use a full factorial design;
— If changing the factor is expensive in terms of cost or use of resources, the changes need to be minimized: use fractional or block design.

---

4   For the interested reader to learn more about these statistical design of experiments, some references are given at the end of this chapter.

8

● **Table 8.7** Optimal combinatorial library based on design of experiments

res	ori	prom1	PAL	prom2	4CL	prom3	CHS	CHI	Chassis	Media
amp	psC101	pBAD	P45735	Ptrc	O24540	pBAD	P48388	P28012	DH10b	LB
kan	pBR322	pBAD	P45726	pBAD	P41636	Ptrc	P48387	P28012	DH5a	M9
kan	psC101	Ptrc	P45726	Ptrc	P41636	pBAD	Q8RVK9	P11650	DH5a	LB
kan	psC101	pBAD	P45727	pBAD	P41636	pBAD	P48388	P41088	DH10b	LB
amp	pBR322	pBAD	P35510	no	P41636	no	Q8RVK9	P28012	MG1655	LB
kan	pBR322	Ptrc	P45735	pBAD	P31686	Ptrc	Q8RVK9	P41088	DH5a	LB
kan	pBR322	Ptrc	P35510	pBAD	O24145	pBAD	P48388	P11650	MG1655	M9
amp	pBR322	Ptrc	P45735	pBAD	P41636	no	P51090	P11650	DH10b	TB
amp	pBR322	Ptrc	P35510	Ptrc	O24540	Ptrc	P51090	P41088	DH5a	LB
amp	psC101	pBAD	P35510	pBAD	O24145	Ptrc	Q8RVK9	Q93XE6	DH10b	M9
kan	pBR322	Ptrc	P35510	Ptrc	P31686	pBAD	P48387	Q93XE6	DH10b	TB
kan	psC101	Ptrc	P45727	no	P31686	Ptrc	P51090	P28012	DH10b	M9
amp	psC101	Ptrc	P45727	pBAD	O24145	pBAD	P48388	P28012	DH5a	TB
kan	psC101	Ptrc	P45726	pBAD	O24540	no	P48387	Q93XE6	MG1655	LB

amp	pBR322	pBAD	P45727	Ptrc	P31686	no	P48388	Q93XE6	DH5a	M9
amp	psC101	Ptrc	P45735	no	P41636	Ptrc	P48388	Q93XE6	MG1655	TB
kan	pBR322	pBAD	P45735	no	O24145	pBAD	P51090	Q93XE6	DH5a	LB
kan	pBR322	pBAD	P45727	Ptrc	O24540	Ptrc	Q8RVK9	P11650	MG1655	TB
kan	psC101	pBAD	P45735	Ptrc	O24145	no	P48387	P41088	MG1655	M9
amp	psC101	pBAD	P45726	pBAD	P31686	pBAD	P51090	P41088	MG1655	TB
amp	pBR322	Ptrc	P45726	no	O24540	pBAD	Q8RVK9	P41088	DH10b	M9
kan	pBR322	Ptrc	P45726	Ptrc	O24145	no	P48388	P28012	DH10b	TB
kan	psC101	pBAD	P35510	no	O24540	no	P48388	P11650	DH5a	TB
amp	pBR322	pBAD	P45726	no	O24145	Ptrc	P48387	P11650	DH10b	LB

8.5   Pathway Standardization

Once we have selected the genetic parts that form the building blocks of the pathway, we need to put all the pieces together and generate the required information so that the DNA sequences of each of them can be designed, synthesized and the resulting pathways assembled. This process has been traditionally performed through different approaches involving manual supervision. There is nowadays a consensus in the community that the **use of a standard representation** is needed in order to optimize the DNA design process and automate the pathway build. An engineering solution that is being developed for that purpose is the Synthetic Biology Open Language (SBOL) [1]. SBOL has been developed in order to guarantee the interoperability of tools and standards used in engineering biology. Moreover, SBOL leverages ontologies, such as gene ontology or systems biology ontology. SBOL consists of two standards, a data model as a formalized representation of data objects and a standardized set of visual schematic symbols.

There are several data repositories available, including ICE[5] [2] and SynBioHub[6] [4] as well as sequence editors such as SBOL Designer [5]. We will see an example of a possible implementation of our pathways using SBOL in the following by using pysbol,[7] the Python library for SBOL. The full details of the SBOL specifications are beyond the scope of this textbook but they can be accessed through the online documentation.[8]

SBOL encourages reusing DNA parts. In the following example, we build a basic construct composed of an inducible promoter, ribosome binding site (RBS), enzyme gene sequence and terminator. In the SynBioHub repository there are many DNA parts coming from different sources such as the ones that each year are generated by the iGEM competition (see ▶ Box 8.3) as well as those coming from synthetic biology institutes and biofoundries like JBEI[9] and SYNBIOCHEM.[10] We will select the promoter from a well-known collection of constitutive and inducible promoters that were developed at JBEI [3]. At SynBioHub, we can browse the collections from the JBEI Public Registry. For our example, we selected the collection by Lee et al. 2011 [3]. Among the long list of parts that are listed, we selected ▶ https://synbiohub.org/public/jbei/JPUB_000111/current, which is a BioBrick (see ▶ Box 8.4) that contains the Tet promoter and tetR tetracycline resistance protein, the RBS and a green fluorescent protein (GFP) gene. We will reuse the promoter and RBS in our construct (◘ Fig. 8.2).

Box 8.3

The **International Genetically Engineered Machine (iGEM)** (▶ http://igem.org) competition is a worlwide synthetic biology competition for undergraduate university students. In recent years, it has expanded to high school students, entrepreneurs and other synthetic biology enthusiast communities. iGEM has been a key actor in order to introduce synthetic biology concepts to students and increase its outreach. iGEM provides a increase large collection of legacy DNA parts and functional devices that can be reused by other groups. iGEM has also contributed to synthetic biology awareness for funding bodies, investors and general public.

---

5   ▶ https://github.com/JBEI/ice
6   ▶ https://synbiohub.org/
7   Note that at the time of this writing pysbol is only available for Python 2.7 and Python 3.6.
8   ▶ https://github.com/SynBioDex
9   ▶ http://jbei.org
10 ▶ http://synbiochem.co.uk/

> **Box 8.4**
> **BioBrick** parts are DNA sequences that conform to an assembly standard. In that way, the output of an assembly transformation can be used as the input to any subsequent manipulation [7]. A BioBrick part needs to be constructed only once in order to be added to the library of parts that can be later reused by other synthetic biologists and in more complex assemblies. The most widely used BioBrick assembly standard involves the use of restriction enzymes. The DNA sequence of each BioBrick part is carried by a circular plasmid, which acts a vector. The BioBrick Foundation (BBF) (▶ https://biobricks.org) is a public-benefit organization established to promote the use of standardized BioBricks parts.

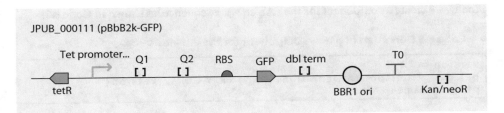

JPUB_000111 (pBbB2k-GFP)

■ **Fig. 8.2** GFP expression with Tet inducible promoter. Diagram generated by `SynBioHub`

In the following Code 8.1, we will:

1. Create a new SBOL document;
2. Pull from the `SynBioHub` repository the BioBrick containing the promoter and RBS;
3. Extract the components associated with the promoter and RBS.

■ ■ **Code 8.1 Creation of a SBOL document and retrieval of promoter and RBS parts from a BioBrick.**

```
from sbol import *
# Set a unique default namespace
namespace ="http://metpathdes.org"
setHomespace(namespace)
# Create a new SBOL document
doc = Document()
# Start an interface to igem s public part shop on SynBioHub.
partshop = PartShop('https://synbiohub.org/public/igem')
# Pull the BioBrick from JBEI for the GFP plasmid
partshop.pull(' https://synbiohub. org/public/jbei/JPUB _000111/
    current', doc)
# Loop through the sequence annotations and extract the promoter
    and RBS
jpub = doc.componentDefinitions['https://synbiohub.org/public/
    jbei/JPUB000111/1']
for c in jpub.sequenceAnnotations:
    comp = c.extract()
    if comp.name == 'Tet_promoter_region':
        ptrc = comp
        ptrc.name = 'Tet_promoter_region'
    elif comp.name == 'RBS ':
        rbs = comp
        rbs.name = 'RBS '
```

For the terminator, we will take the BioBrick ▶ https://synbiohub.org/public/igem/ BBa_B0010/1 from the iGEM library (Code 8.2).

■■ **Code 8.2 Retrieval of a terminator component from the iGEM library.**

```
# Terminator
partshop.pull('http://synbiohub.org/public/igem/BBa_B0010/1',
    doc)
term = doc.componentDefinitions['https://synbiohub.org/public/
    igem/BBa_B0010/1']
term.name = 'Terminator'
```

Finally, we create a component for the PAL enzyme sequence, as shown in Code 8.3.

■■ **Code 8.3 Definition of a new component for the enzyme gene.**

```
version = "1.0.0"
pal_cds = ComponentDefinition("PAL", BIOPAX_DNA, version)
pal_cds.name = "PAL"
```

As an example, we will provide the sequence from the *Petroselinum crispum* (parsley) source, with UniProt identifier P24481, which will be downloaded from the European Bioinformatics Institute (EBI) by following the steps in the Code 8.4.

■■ **Code 8.4 Retrieval of PAL sequence from EBI.**

```
import requests
url = 'https://www.ebi.ac.uk/ena/data/view/CAA68938&display=fasta'

r = requests.get(url)
fasta = r.text.split('\n')
P24481 = ''.join(fasta[1:])
palp_seq = Sequence("PAL_Parsley_seq", P24481,
    SBOL_ENCODING_IUPAC, version)
# Add the sequence to the SBOL document
doc.addSequence( palp_seq )
# Add the PAL component to the SBOL document
pal_cds.roles = SO_CDS
pal_cds.sequences = palp_seq
doc.addComponentDefinition(pal_cds)
```

We have now all the elements in the document and it is time now to assemble them together by creating a new component, which will the new construct, as shown in Code 8.5.

■■ **Code 8.5 Create and assembly a full construct.**

```
# Create a new empty device named 'my_device'# Creat
my_device = doc.componentDefinitions.create('my_construct')
my_device.name = "my_construct"
# Assemble the new device from the promoter, rbs, cds, and
    terminator from above.
my_device.assemblePrimaryStructure(
    [ptrc, rbs, pal_cds, term])
# Set the role of the device with the Sequence Ontology term
    'gene'
my_device.roles = SO_GENE
# Compile the sequence for the new device
my_device.compile()
```

The resulting construct can be saved as an SBOL document that can be exchanged with other software, collaborators or DNA vendors. In addition, the new construct will be submitted to a private area of SynBioHub. To that end, you will need to create a personal account on SynBioHub in order to provide your account details, as shown in Code 8.6. The resulting uploaded construct is shown in ■ Fig. 8.3.

■■ **Code 8.6 Submission of the construct to SynBioHub.**

```
# Connect to your account on SynBioHub
import getpass
email = getpass.getpass()
password = getpass.getpass()
partshop.login(email, password)
# Give your document a displayId, name, and description
doc.displayId = 'my_construct'
doc.name = "Construct"
doc.description = "An example of a simple construct"
# Submit the document to the part shop
partshop.submit(doc)
```

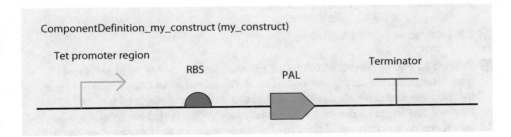

ComponentDefinition_my_construct (my_construct)

Tet promoter region

RBS

PAL

Terminator

■ **Fig. 8.3**   Resulting promoter+RBS+PAL enzyme construct. Diagram generated by SynBioHub

┌───┐
  Take Home Message

  - Pathway design involves the selection of the enzyme sequences based on multiple criteria.
  - Transcriptional and translational tuning involve other design decisions.
  - Additional design parameters are origin of replication, resistance cassette, chassis and growth media.
  - Statistical design of experiments allows narrowing down the number of combinations that need to be tested.
  - Standard representation and public deposit of the resulting constructs can be achieved by using the Synthetic Biology Open Language (SBOL).
└───┘

8.6 Problems

8.1 Alternative enzyme selection.

Tables 8.1, 8.2, 8.3, and 8.4 provide a possible selection of enzyme sequences based on the described criteria. Go to the `Selenzyme`[11] tool, run a search that outputs the top 200 sequence candidates and perform alternative selections based solely on one of the available criterium as follows:
1. Reaction similarity;
2. Taxonomic distance;
3. Molecular weight.

Which criteria do you think would be the most appropriate? Do you think that there is an *optimal criteria* for all the enzymes or that it should be decided on a case-by-case basis?

8.2 An extended library of synthetic vectors.

Lee and colleagues [3] developed and characterized a full library of vectors with different resistance cassettes, origin of replications and inducible promoters (see for instance ◻ Figure 2 in Ref. [3]). If we used this same full library for our pathway design based (◻ Table 8.6), how many total combinations will contain the design space? *Hint:* the promoter choices should be increased in all cases.

8.3 Alternative experimental designs.

◻ Table 8.7 can be generated using DoE software such as `JMP` or `Design Expert` where the size of the library was limited to 24. Download a free trial version of one of this software and generate a custom design of a library consisting of 24, 48 and 72 different constructs for the extended design space defined in previous Question 8.2. Which one do you think would be the most appropriate choice for the extended library?

? 8.4 SBOL-based library design.

✓ Generate a combinatorial library such as the one shown in ▪ Table 8.7 of 10 constructs using SBOL. Upload the resulting library to `SynBioHub`. What is the coverage of your library compared with the total design space (total number of combinations)?

References

1. Cox, R.S., Madsen, C., McLaughlin, J.A., Nguyen, T., Roehner, N., Bartley, B., Beal, J., Bissell, M., Choi, K., Clancy, K., Grünberg, R., Macklin, C., Misirli, G., Oberortner, E., Pocock, M., Samineni, M., Zhang, M., Zhang, Z., Zundel, Z., Gennari, J.H., Myers, C., Sauro, H., Wipat, A.: Synthetic Biology Open Language (SBOL) version 2.2.0. J. Integr. Bioinform. **15**(1) (2018). https://doi.org/10.1515/jib-2018-0001
2. Ham, T.S., Dmytriv, Z., Plahar, H., Chen, J., Hillson, N.J., Keasling, J.D.: Design, implementation and practice of JBEI-ICE: an open source biological part registry platform and tools. Nucl. Acids Res. **40**(18), e141–e141 (2012). https://doi.org/10.1093/nar/gks531
3. Lee, T., Krupa, R.A., Zhang, F., Hajimorad, M., Holtz, W.J., Prasad, N., Lee, S., Keasling, J.D.: BglBrick vectors and datasheets: a synthetic biology platform for gene expression. J. Biolog. Eng. **5**(1), 12 (2011). https://doi.org/10.1186/1754-1611-5-12
4. McLaughlin, J.A., Myers, C.J., Zundel, Z., Misirli, G., Zhang, M., Ofiteru, I.D., Goñi Moreno, A., Wipat, A.: SynBioHub: a standards-enabled design repository for synthetic biology. ACS Synth. Biol. acssynbio.7b00403 (2018). https://doi.org/10.1021/acssynbio.7b00403
5. Quinn, J.Y., Cox, R.S., Adler, A., Beal, J., Bhatia, S., Cai, Y., Chen, J., Clancy, K., Galdzicki, M., Hillson, N.J., Le Novère, N., Maheshwari, A.J., McLaughlin, J.A., Myers, C.J., P, U., Pocock, M., Rodriguez, C., Soldatova, L., Stan, G.B.V., Swainston, N., Wipat, A., Sauro, H.M.: SBOL visual: a graphical language for genetic designs. PLOS Biol. **13**(12), e1002,310 (2015). https://doi.org/10.1371/journal.pbio.1002310
6. Salis, H.M.: The ribosome binding site calculator. In: Methods in Enzymology, vol. 498, pp. 19–42. Elsevier (2011). https://doi.org/10.1016/B978-0-12-385120-8.00002-4
7. Shetty, R.P., Endy, D., Knight, T.F. Jr.: Engineering BioBrick vectors from BioBrick parts. J. Biolog. Eng. **2**, 5 (2008). https://doi.org/10.1186/1754-1611-2-5

Further Reading

A more detailed discussion focused on **parameter tuning** in genetic designs:

Arpino, J.A.J., Hancock, E.J., Anderson, J., Barahona, M., Stan, G.B.V., Papachristodoulou, A., Polizzi, K.: Tuning the dials of Synthetic Biology. Microbiology **159**(Pt_7), 1236–1253 (2013). https://doi.org/10.1099/mic.0.067975-0
Jones, J., Koffas, M.: Optimizing metabolic pathways for the improved production of natural products. Methods Enzymol. **575**, 179–193 (2016). https://doi.org/10.1016/BS.MIE.2016.02.010.

An excellent and comprehensive reference for **statistical design of experiments**:

Montgomery, D.C.: Design and analysis of experiments. John Wiley & Sons (2017).
The Design of Experiments Guide of the `JMP` software (https://www.jmp.com/) is also a useful source of information about DoE, illustrated with multiple examples based on the use of the software.

Pathway Redesign

© Springer Nature Switzerland AG 2019
P. Carbonell, *Metabolic Pathway Design*, Learning Materials in Biosciences,
https://doi.org/10.1007/978-3-030-29865-4_9

What You Will Learn in This Chapter

Previous chapters have focused on metabolic pathway and their genetic parts modeling and selection. In this chapter, we direct our interest towards pathway design in order to move into the engineering aspects of metabolic pathway design. Our focus from now on will not be on reproducing through simulation some biological behavior or trying to understand its mechanism but in determining the possible biological interventions and modifications to be engineered in the cell in order to achieve the desired behavior. Genetic parts selection and experimental design will be essential at this stage. Since this book is about metabolic pathway design, metabolism will continue as our central topic. However, we will also consider solutions that go beyond metabolism and will take advantage of the advanced capabilities that synthetic biology provides to metabolic engineering.

9.1 Model-Based Media and Chassis Redesign

A truly optimized engineered pathway should be able to work under all required different conditions. For instance, an always desirable feature of any metabolic pathway design is to select the best growth media environment that allows the best response in the strain. In order to improve the performance of the strain, the metabolism of the chassis can be modified by knocking-down those genes that have an impact on important reactions in pathway-related responses. Genome-scale models allow us approaching this problem in a systematic way.

Because of the complexity of metabolic networks, we cannot always expect predictions to be completely accurate or to cover all possibilities, but at least models should be able to provide interesting hints about what types of **redesigns** could be introduced into the media or the chassis in order to gain some improvements in the performance objectives.

We start by engineering a model (adding new metabolites and reactions) based on the default *E. coli* model in `cobrapy` introduced in ▶ Sect. 2.2 with our target flavonoid pathway as in Code 9.1.

■ ■ **Code 9.1 Definition of an *E. coli* model with an engineered flavonoids production pathway.**

```
import cobra
import cobra.test

model = cobra.test.create_test_model("ecoli")

# Define pathway heterologous metabolites
pino = cobra.Metabolite('pino_c', compartment='c',
                                        name='pinocembrin
                                            ')
pinocha = cobra.Metabolite('chal_c', compartment='c',
                                        name='pinocembrin
                                            -chalcone')
maloncoa = model.metabolites.get_by_id("malcoa_c")
cinnacoa = cobra.Metabolite('cinn_c', compartment='c',
                                        name='cinnamoyl-
                                            coa')
```

```
# Retrieve pathway metabolites already defined in the model
coa = model.metabolites.get_by_id( "coa_c")
cinn = model.metabolites.get_by_id( "cinnm_c")
co2 = model.metabolites.get_by_id( "co2_c")
atp = model.metabolites.get_by_id( "atp_c")
amp = model.metabolites.get_by_id( "amp_c")
ppi = model.metabolites.get_by_id( "ppi_c")
phen = model.metabolites.get_by_id( "phe__L_c")
nh4 = model.metabolites.get_by_id( "nh4_c")

# Define heterologous reactions in the pathway
pal = cobra.Reaction('PAL ')
pal.add_metabolites({ phen: -1.0, cinn: 1.0, nh4: 1.0})
r4cl = cobra.Reaction('4CL ')
r4cl.add_metabolites({ cinn: -1.0, atp: -1.0, coa: -1.0,
                                cinnacoa: 1.0, amp: 1.0,
                                ppi: 1.0})

chs = cobra.Reaction('CHS ')
chs.add_metabolites({ maloncoa: -3.0, cinnacoa: -1.0,
                                pinocha: 1.0, co2: 3.0,
                                coa: 4.0})

chi = cobra.Reaction('CHI ')
chi.add_metabolites({ pinocha: -1.0, pino: 1.0})
rt = cobra.Reaction('RT ')
rt.add_metabolites({ pino: -1.0 })
model.add_reactions([pal, r4cl, chs, chi, rt])
for r in [pal, r4cl, chs, chi, rt]:
    r.lower_bound = 0
    r.upper_bound = 1000
```

In this example, we will make the following assumptions about the environmental conditions:

- Variable availability of the carbon source (D-glucose);
- Depletion of glucose is supplemented in the growth medium by L-phenylalanine precursors;
- Variable aerobic conditions (O_2 availability).

Such simple scenarios are useful in order to understand how the carbon source, supplemental feeding or aerobic conditions can affect the performance of the strain. As we already discussed in ▶ Sect. 7.2, performance of the engineered strain will be assessed through production capabilities based on coupled yield-growth. The objective function will be defined by the **product of the maximum achievable production of the target compound and the maximum biomass** and it will be computed for several environmental conditions. As shown in Code 9.2, we can run simulations for the engineered model of Code 9.1 for a grid of different values by limiting the maximum achievable exchange fluxes of glucose and oxygen.

■■ **Code 9.2 Media screening for the engineered strain.**

```
biomass = model.reactions.get_by_id('
    Ec_biomass_iJO1366_core_53p95M')
rt = model.reactions.get_by_id('RT')

# Change bounds
def change_bounds(cobra_model, reaction_id, lb, ub):
    reac = cobra_model.reactions.get_by_id(reaction_id)
    reac.lower_bound = lb
    reac.upper_bound = ub

# Simulate different enviromental conditions
def simulate(model, o2, phe, glc):
    change_bounds(model, 'EX_o2_e', -o2, 1000)
    change_bounds(model, 'EX_phe_L_e', -phe, 1000)
    change_bounds(model, 'EX_glc_e', -glc, 1000)
    try:
        solution = model.optimize()
        val = solution.objective_value
    except:
        val = 0
    return val
data = []
for o2 in np.arange(5,100,5):
    for glc in np.arange(5,100,5):
        phe = (100-glc)
        model.objective = biomass
        biom = simulate(model, o2, phe, glc)
        model.objective = rt
        rtf = simulate(model, o2, phe, glc)
        print(o2,glc,phe,biom*rtf)
        data.append( np.array( (o2,glc,biom, rtf, biom*rtf) ))

data = np.array( data )
```

The results of the simulations can be plotted using the `matplotlib` (see Appendix A) plotting library as shown in ◨ Fig. 9.1 by running Code 9.3.

■■ **Code 9.3 Media screening evaluation of strain performance.**

```
import matplotlib.pyplot as plt
%matplotlib inline
from mpl_toolkits.mplot3d import Axes3D
fig = plt.figure()
ax = fig.add_subplot(111, projection='3d',
                             xlabel='$O_2$',
                             ylabel='$D$-Glucose',
                             zlabel='Strain_
                                 performance')
ax.scatter(xs=data[:,0],ys=data[:,1],zs=data[:,4])
```

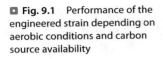

■ **Fig. 9.1** Performance of the engineered strain depending on aerobic conditions and carbon source availability

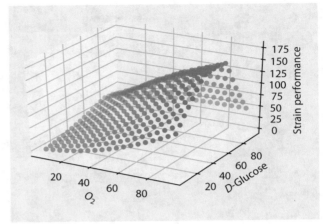

According to the simulations, the resulting optimal environmental conditions were found to be at 95 for O_2 exchange flux (full aerobic), with availability of glucose of 45. The equilibrium of the fluxes can be then computed by setting up these optimal limits in the function `simulate` defined in Code 9.2 as shown in the following Code 9.4.

■ ■ **Code 9.4 Simulation of the engineered strain at optimal growth conditions.**

```
simulate(model, 95, 45, 55)
model.optimize()
```

Besides simulating environmental conditions for growth, another important approach for model-based redesign is **to simulate modifications on the chassis** by introducing gene knock-outs or deletions. The effect of deleting a gene that encodes some enzyme will generally lead to the blockage of some reaction or set of reactions in the metabolic network. In the case where more than one gene encodes an enzyme for the reaction, a single knock-out would still have some impact but it will not completely shutdown the catalytic conversion. In `cobrapy` we have some routines that allow us to simulate the effect of single and double gene deletions on cell's objective function.

Code 9.5 shows an example of a single knock-out simulation for the full set of genes in the model. The model that is used in this example is a simplified model for *E. coli*, generally known as the **E. coli core model**. The reason for using a simplified version of the metabolic network is that simulations for gene deletions can be computationally expensive because they have to run flux balance analysis simulations multiple times. Therefore, it is often necessary to perform some simplifications in the models or to focus on a subset of target knock-out genes.

■ Table 9.1 lists some of the genes in the model that led to the highest positive change in growth after running the single gene deletion of Code 9.5. As can be seen from the table, there is no single deletion providing the highest increase in biomass in the model but there is a set of genes each one having a positive impact on growth. In order to look for the best

◘ **Table 9.1**   Top predicted gene knock-outs

ids	Growth	Status
(b0733)	0.87	Optimal
(b3925)	0.87	Optimal
(b1676)	0.87	Optimal
(b3603)	0.87	Optimal
(b3213)	0.87	Optimal
(b4122)	0.87	Optimal
(b3612)	0.87	Optimal
(b1276)	0.87	Optimal
(b1723)	0.87	Optimal
(b2464)	0.87	Optimal
(b2935)	0.87	Optimal
(b0902)	0.87	Optimal
…	…	…

candidate genes we can go further and try with double deletions. However, computing all pairs of double mutants would take a long time (probably more than 24 h) and therefore Code 9.6 is limited to the computation of the pairs between the top 5 genes in the list of single mutants. The result of performing the double mutant simulations is shown in ◘ Table 9.2. As we can see on the table, there is no obvious increase on biomass from any pair of mutants. However, some of them led to a decrease in the total biomass, i.e., they had a cooperative negative effect. Such type of cooperative negative effects are often found, so that genes that individually had a positive impact when knocked-out had a negative effect when knocked-down simultaneously, arriving in some cases to total inhibition of growth (what is generally known as *synthetic lethality*).

▪▪ **Code 9.5 Single gene deletion.**

```
from cobra.flux_analysis import double_gene_deletion,
    single_gene_deletion
model = cobra.test.create_test_model("textbook")
gene_deletion = single_gene_deletion(model)
gene_deletion = gene_deletion.sort_values('growth',
                            ascending=False)
display( gene_deletion )
```

□ Table 9.2 Top predicted double gene knock-outs

ids	Growth	Status
(b0351, b1849)	0.87	Optimal
(b3115, b1849)	0.87	Optimal
(b0351, b1241)	0.87	Optimal
(b3115, b1241)	0.87	Optimal
(b0351)	0.87	Optimal
(b1849, b1241)	0.87	Optimal
(b3115)	0.87	Optimal
(b1849)	0.87	Optimal
(b1241)	0.87	Optimal
(b0351, b3115)	0.87	Optimal
(s0001)	0.21	Optimal
(b0351, s0001)	0.21	Optimal
(b3115, s0001)	0.21	Optimal
(b1849, s0001)	0.21	Optimal
(b1241, s0001)	0.21	Optimal

■ ■ Code 9.6 Double gene deletion.

```
gene_deletion = gene_deletion.sort_values('growth',
                              ascending=False)
# Sort genes based on their effect on growth
rl1 = [list(i)[0] for i in gene_deletion.index ]
db = double_gene_deletion(model, model.genes[0:5],
                number_of_processes=4,
                return_frame=False).round(4)
display( db )
```

9.2 Computational Enzyme Redesign

Rather than introducing mutants that completely knock-down genes in the metabolic network, as described in the previous ▶ Sect. 9.1, a more subtle but also effective way of performing the pathway redesign is through **computational enzyme design** [14] (see ▶ Box 9.1). The golden rule of computational enzyme design is to identify the regions in the protein that

> **Box 9.1**
>
> **Enzyme engineering** is a process that searches for mutated variants of the parent protein with improved properties. Strategies for enzyme engineering include:
> - Rational design (mutagenesis), which relies on biophysical and biochemical knowledge to predict specific mutations improving the desired activity;
> - Random selection (**directed evolution**), which circumvents the challenge of predicting the best mutations by applying an assay that can test many of them. Directed evolution (see also ► Box 5.1) consists of consecutive rounds of error-prone polymerase chain reaction (PCR) and DNA shuffling followed by screening and selection.
>
> Enzyme engineering generally requires of high-throughput screening, especially for directed evolution, making the process challenging from a cost-effectiveness point of view. Computational enzyme redesign can provide *in silico* protocols that combine physico-chemical knowledge from biological databases with scoring functions from bioinformatics modeling to optimally select candidate solutions.

are functionally linked to the enzyme activity and then to introduce some substitutions (mutations) that enhance its desired activity. There are several steps involved in this process:

1. Similarity search: as discussed in ► Chap. 5 this can be performed based on sequence, structure, ligand or reaction;
2. Identification of functional regions or hot-spots in the protein;
3. Finding the list of candidate mutants that could improve the enzyme activity;
4. Combinatorial *in silico* screening of the mutants.

In order to identify the candidate mutants, two approaches are possible in computational enzyme redesign:

1. **Knowledge-based approaches** consist of collecting all relevant information about the enzyme class or family from sequence, structure and activity information in databases in order to predict the best mutants;
2. **Energy-based approaches** consist of performing a simulation based on the biophysical modeling of molecular interactions and dynamics in order to predict the best mutants.

As in ► Chap. 5, **multiple sequence alignments** within proteins of the same family are going to be a valuable source of information in the **knowledge-based approach** to select mutant candidates. Either a homology search based on BLAST or a similarity measure based on motif or pattern content will help us to identify closer sequences. As we saw in ► Sect. 5.1, enzymes can be classified in multiple ways (see ► Box 5.2). For instance:

- ENZYME[1] database from ExPASy provides multiple links to protein information regarding the sequences for the given EC class (enzyme classification number according to the Enzyme commission);
- IntEnz,[2] which provides the link to annotated sequences in UniProt, which can be reduced into clusters by filtering redundancy (a tool like CD-HIT [9] can conveniently perform such task for a sequence list);
- PROSITE,[3] in turn, provides the link to proteins containing the pattern or signature of the enzyme family.

◼ **Table 9.3** Selection of online tools for the identification of protein functional sites and hot-spots

Name	Method
Consurf	Conservation of PDB structures
	▶ http://consurf.tau.ac.il
CASTP	Calculation of pockets and cavities
	▶ http://sts.bioe.uic.edu/castp
CSA	Catalytic site atlas
	▶ http://www.ebi.ac.uk/thornton-srv/databases/CSA/
PredictProtein	Multiple predictors of protein structure and function
	▶ https://www.predictprotein.org/
WHAT IF	Multiple predictors of protein structure and function
	▶ https://swift.cmbi.umcn.nl/whatif/

A multiple sequence alignment of enzymes in the same class can then provide information about residues where mutations will bring specificity towards each substrate. In general, conserved regions are related to some structural role, while regions showing variability in parallel with multiple substrates might be related to positions determining substrate affinity.

Similarly, another important source of information is found in **protein structures** such as the ones available in the Protein Data Bank (PDB).[4] Structural information provides another perspective helping on identifying potential functional sites for mutations:

— As structure is more conserved than sequence, multiple sequence alignments deduced from structure superposition are often more precise;
— Visual inspection of mutations around the catalytic site could provide valuable clues for possible candidate substitutions;
— Protein structure superposition like the ones performed by MAMMOTH [12] or Dali [7], as discussed in ▶ Sect. 5.1, could help at prioritizing mutants. For example by discriminating buried residues versus those that are solvent accessible or by identifying residues that are in close contact with the region of interest of the protein.

Selection of potential mutants can be done either by inspecting the sequence alignment alone or by introducing some additional information from databases such as Brenda,[5] PubChem Bioassay[6] or protein interaction databases like IntAct.[7] In order to identify potential regions containing candidate mutants, many bioinformatics tools are available allowing the identification of functional sites of the protein from sequence or structure. A selection of some of the most important tools is listed in ◼ Table 9.3. Some are

---

4 ▶ https://www.wwpdb.org/
5 ▶ https://www.brenda-enzymes.org
6 ▶ https://pubchem.ncbi.nlm.nih.gov
7 ▶ https://www.ebi.ac.uk/intact/

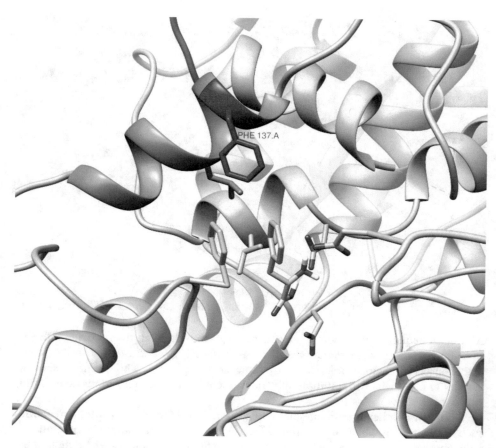

◻ Fig. 9.2 Docking of *L*-phenylalanine to PAL (PDB id 1W27) by using `1-click docking`. In red, hot-spot residue F137, in green the α-helix containing F137

based on conservation of structures or sequences. Many others look for pockets or cavities in the protein through the analysis of structural and geometric properties.

For instance, for the PAL enzyme, we selected in ▶ Sect. 5.2 the sequence P24481 from `UniProt`, which has an associated 3D structure 1W27 in the `PDB` database. According to the Catalytic Site Atlas (`CSA`), residues Tyr110, Tyr351 and Phe400 might have a functional role. Moreover, `Brenda` reports that mutations F137I, F137T, and F137V had a positive increase in the PAL activity compared the wild-type enzyme [2]. This can be investigated further by using some of the online docking servers to predict the interaction between the protein and the ligand (in this case between the enzyme and the substrate) such as `SwissDock`,[8] `1-click docking`[9] and many others. ◻ Figure 9.2 shows a graphical rendering from `Chimera` of the top ranked spatial pose from docking of *L*-phenylalanine into the protein structure 1W27 of the PAL enzmye using `1-click docking`. In order to perform the docking modeling, the binding site center was selected in the region of F137.

---

8   ▶ http://www.swissdock.ch/
9   ▶ https://mcule.com/apps/1-click-docking/

Once the **region of interest** (functional, active, catalytic site, etc.) of the protein have been identified, **energy-based methods** aim at predicting the effect of mutations from first principles[10] from molecular and quantum mechanics. **Molecular dynamics** generally helps accelerating the computation while **quantum mechanics** can provide higher accuracy. Conformational sampling of the protein is usually done using force fields implemented into Monte-Carlo or molecular dynamics procedures. To estimate the influence of a mutant side-chain at the protein binding interface the number of degrees of freedom can be restricted to the region of interest.

For a given choice of side-chains, we search for the most favorable configuration in terms of **energetic fitness** among the allowed interacting positions of the substrate on the enzyme in order to model the transition state of substrate-enzyme. The cluster of substrate conformers with the lowest fitness energy is estimated. Fitness considers internal, solvent and surface energies. The interface is then computed in order to determine residues that belong to the catalytic site.

Several software packages exist that perform energy calculations for computational protein design, such as FoldX[11] [3] or Rosetta [13]. For instance, the FRESCO [17] protocol employs a combination of FoldX and Rosetta in order to design mutant libraries improving protein stability. In the Rosetta method, protein backbone conformation is specified as a list of backbone torsion angles, while side chains are restricted to discrete conformations from a backbone-dependent rotamer library. Many Rosetta-based computational protocols for protein design are available.[12] For instance a protocol for enzyme design is available that models the reaction transition state using quantum chemistry methods in order to define side chain and backbone constrains associated with positions optimal for catalysis.

Because of the large diversity of computational protein design software and approaches that are available, selecting the most appropriate method in each case can be challenging. In order to alleviate the decision process, several automated design protocols for libraries of enzyme mutants are available. For instance, FuncLib[13] [8] is an online method that designs a set of diverse multipoint or multiple mutants in enzyme active sites. Screening for multipoint mutations is important because of cooperativity effects, i.e., single mutants might have an effect that is completely different to the simultaneous effect of multiple mutation (mutational epistasis).

In order to test the enzyme design capabilities of Funclib the 3D structure of the PAL enzyme with PDB id 1W27 was submitted to the server. Since the number of positions that can be mutated is limited, we selected the positions around the α-helix found at residues A129 to N139N, which includes the position F137F that has been identified in the literature as hot-spot, as shown in ◘ Fig. 9.2. The mutant library was limited from 2 up to 4 simultaneous mutations. The designated substitutions were calculated by the server as shown in ◘ Fig. 9.3 totaling 79,525 possible combinations. The top predicted mutants according to the energy score that were obtained from such calculation, which took several hours to complete in the server, are listed in ◘ Table 9.4. Interestingly, top mutations (those with the lowest energies) appeared clustered at positions L130, A132, and F137 (mutations at both ends of the helix did not seem to follow a pattern and therefore can be ignored). This was consistent with the information reported by the Brenda data-

---

10 A principle based on the established laws of physics, i.e., no empirical or fitting parameters.
11 ► http://foldxsuite.crg.eu/
12 ► https://www.rosettacommons.org
13 ► http://funclib.weizmann.ac.il/

base on the positive effects of mutations F137I and F137V [2]. According to the predictions with the lowest energies in ▢ Table 9.4, F137I/L130V, F137I/K132R and F137V/K132R can be considered as promising double mutant candidates with favorable impact on PAL activity.

Parameter	Value
Minimal number of mutations per design	2
Maximal number of mutations per design	4
Minimal PSSM threshold	-1
ΔΔG	3.0
Difference between clustered variants	2
Sequence space	

129A	ADEGHKLMNQRST
130A	LV
131A	Q
132A	KADEFHILMNQRSTVY
133A	EIQ
134A	LFI
135A	IMV
136A	RAEHKSTV
137A	FIMV
138A	L
139A	NEGHQST

Total number of designs in tolerated sequence space	79,525

▢ **Fig. 9.3** Configuration of running parameters for the `FuncLib` server

▢ **Table 9.4** Best mutants and scores predicted by `FuncLib` from the parent sequence (first row) focused on the α-helix region of residues 129–139, chain A of 1W27

129A	130A	131A	132A	133A	134A	135A	136A	137A	138A	139A	total_score
A	**L**	**Q**	**K**	**E**	**L**	**I**	**R**	**F**	**L**	**N**	-1910.303
E	V	Q	K	E	L	I	R	I	L	Q	-1922.012
L	L	Q	R	E	L	I	R	V	L	Q	-1919.782
N	L	Q	R	E	L	I	R	F	L	Q	-1919.631
D	L	Q	R	E	L	I	R	F	L	E	-1919.578
L	L	Q	M	E	L	I	R	F	L	Q	-1919.536
N	L	Q	K	E	L	I	R	V	L	Q	-1919.508
K	L	Q	R	E	L	I	R	F	L	T	-1919.504
K	L	Q	R	E	L	I	R	I	L	Q	-1919.436
K	V	Q	K	E	L	I	R	I	L	T	-1919.366

9.3 Experimental Redesign at the Learn Stage

In ▶ Chap. 8, we learned the importance of performing an optimal experimental design in order to efficiently explore the different combinations of design choices by running a relatively small amount of tests. For our example of a flavonoid pathway, a possible experimental design was given in ◪ Table 8.7 containing 24 different constructs. The experimental results of testing such combinatorial library will inform about the impact of factor changes. In the example, 11 factors were varied simultaneously. Obviously the number of possible combinations is very high and therefore the 24 combinations of ◪ Table 8.7, even if optimally selected, would only provide an initial assessment of the influence of the factors. At best, what we can expect is that factors having a highly significant impact will be identified in this initial round so that we can firstly narrow down the number of combinations that need to be tested in successive rounds.

Note that in ◪ Table 8.7 there are no numerical values but only labels like for instance "LB", "M9" and "TB" for media. Such type of variables are known as **categorical variables** (see ▶ Box 8.2). Of course we can also introduce numerical factors if they can be controlled like temperature or the induction time. In that case, what the experimental design will select is the set of values to be tested in the combinations.

In order to provide an example about learning from experimental data, calculations will be simplified to only 3 factors: origin of replication, media and chassis will be considered. The method applied here can be easily extended to the rest of factors. ◪ Table 9.5 shows experimental titers that were measured in the engineered strains after a round of design–build–test. We go now into the **learn** stage by analyzing the relationships between the factors and the obtained titers.

A first analysis that can be performed is to plot the quantified values against the different factors. For that purpose, we will use the `boxplot` (see ▶ Box 9.2) representation from the `pandas` library (see Appendix A), as in Code 9.7. Note that in order to run the code, we need first to define a `pandas` data frame object with the values in ◪ Table 9.5. This can be easily done by first creating the table with a spreadsheet software as a `csv` file, which can be directly read in `pandas`. The resulting plots are shown in ◪ Figures 9.4, 9.5 and 9.6 for origin of replication, chassis and media, respectively. These plots tell us different stories about the effect of each factor. For instance, the effect of the origin in ◪ Fig. 9.4 is not conclusive, titers for strains with origin pBR322 are generally low producers, although there are a significant number of cases that are actually high (the outliers in the plot). For psC101, what we see is that the experimental values are spread out through the range and the plot does not bring any conclusive result. The effect from chassis selection, in turn, is more significant as shown in ◪ Fig. 9.5. It seems quite obvious that strain MG1655 hosted the highest producers, while DH5α and DH10β both led to low producers. Finally, the effect from selecting the growth media is again not very conclusive from the plot in ◪ Fig. 9.6. In principle, TB seems to be the best case but the values are again spread through the full range and therefore this result is not very conclusive.

■■ Code 9.7 Box plots of the factors using a `pandas` dataframe `df` with the experimental values and labels in ◪ Table 9.5.

```
df = pd.read_csv('table.csv')
d1 = df.boxplot('y',by='ori')
d2 = df.boxplot('y',by='Chassis')
d3 = df.boxplot('y',by='Media')
```

◻ Table 9.5 Experimental titers [mg/L] and standard deviation for the constructs in the combinatorial library with chassis, media and origin of replication as factors

Chassis	Media	ori	y [mg/L]	std
DH10b	LB	psC101	28.44	5.77
DH5a	M9	pBR322	22.26	12.41
DH5a	LB	psC101	7.15	7.58
DH10b	LB	psC101	12.96	3.01
MG1655	LB	pBR322	44.22	4.65
DH5a	LB	pBR322	30.45	8.58
MG1655	M9	pBR322	43.81	22.56
DH10b	TB	pBR322	44.10	11.62
DH5a	LB	pBR322	34.44	12.91
DH10b	M9	psC101	11.57	2.30
DH10b	TB	pBR322	28.99	9.73
DH10b	M9	psC101	29.49	6.23
DH5a	TB	psC101	6.85	5.87
MG1655	LB	psC101	34.06	9.19
DH5a	M9	pBR322	38.32	11.29
MG1655	TB	psC101	41.34	15.59
DH5a	LB	pBR322	29.60	3.69
MG1655	TB	pBR322	47.69	8.09
MG1655	M9	psC101	39.81	2.84
MG1655	TB	psC101	42.48	5.70
DH10b	M9	pBR322	22.37	5.73
DH10b	TB	pBR322	38.25	3.99
DH5a	TB	psC101	12.21	6.84
DH10b	LB	pBR322	26.84	13.63

In order to be able to draw conclusions from the analysis of the experimental data, we need to run **statistical tests**. There are many possible statistical tests that can be run, what is important to keep in mind is that each one serves for testing some hypothesis about the relationship between the response variable (in our case the production level of the target compound) and the independent or explanatory variables (in our case the design factors). There are several libraries available in Python in order to perform statistical analyses, as well as many other alternative software including the open-source statistical language R. We will perform here the analysis using the statsmodels

---

Box 9.2 Representation of Experimental Data Using Box Plots

Box plots are used in descriptive statistics to graphically represent numerical data. They provide a quick appraisal of the distribution of the sample. The box represents the interquartile range and contains a horizontal line at the median of the sample. Box plots may also show vertical lines extending from the boxes (whiskers) indicating variability outside the upper and lower quartiles. Outliers, i.e. those that deviate more than some pre-defined threshold like 1.5 from the upper or lower quartile, may be plotted as individual points.

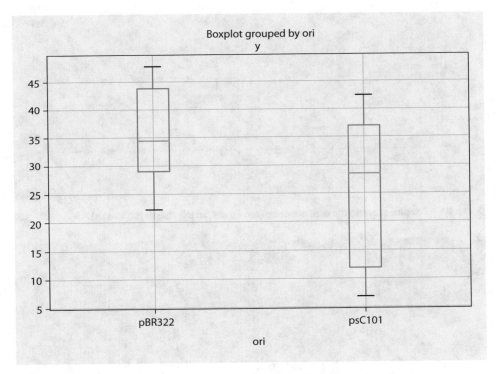

■ **Fig. 9.4** Box plot showing the dependence of the titers y on the origin of replication (ori)

package.[14] Our approach would be to assume that our data can be fitted into a linear model that relates the independent variables to the response. Therefore, we will perform a **regression analysis**. More precisely, we will use *ordinary least squares* (OLS) estimation. The simplest way of performing the linear regression test is by inputting a formula that defines the assumed relationship between the response y and the three factors ori, Chassis, and Media:

$$y \;=\; \beta_0 + \beta_1 \text{ori} + \beta_2 \text{Chassis} + \beta_3 \text{Media} \tag{9.1}$$

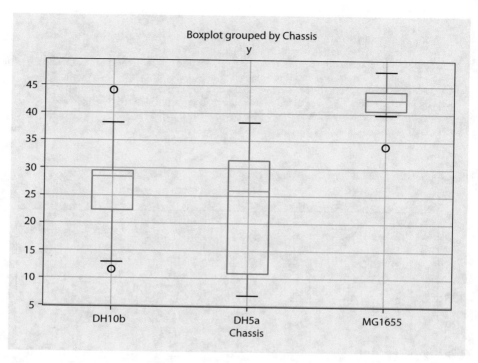

Fig. 9.5 Box plot showing the dependence of the titers y on the chassis (`Chassis`)

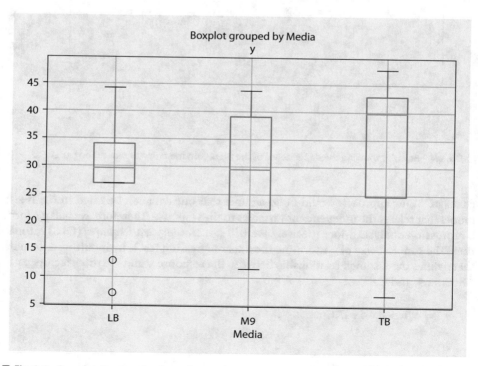

Fig. 9.6 Box plot showing the dependence of the titers y on the media (`Media`)

Nevertheless, as you would have probably realized, there is a problem with Eq. 9.1. The response y is numerical, while the factors are categorical. They are just defined using labels like "LB", "M9" and "TB" without any prior assumption on their actual quantitative effects on the response. Categorical variables are very common in synthetic biology because it is often difficult to have a priori information about the effect of each level on the response. Even if we are using categorical variables, still is possible to perform the regression analysis by converting the categorical variables into quantitative variables. The procedure is quite simple, if the variable contains n levels with their associated labels, $n-1$ binary variables are generated that are set to 1 when the value of the categorical variable is that of the label. The reason for creating only $n-1$ variables is because the n variable is not an independent variable but is determined by the others. In our case, we just need to let statmodels take care of all of these nuisances. As shown in Code 9.8, the formula is input as y~ori+Chassis+Media. This handy way of defining the model is called an R-style formula, since it is often used for that language [15].

■■ Code 9.8 Ordinary linear regression analysis using statsmodels.

```
from matplotlib import pyplot as plt
import statsmodels.formula.api as smf
ols = smf.ols(formula='y~ori+Chassis+Media', data=df)
res = ols.fit()
y = df.loc[:,'y']
yp = res.predict()
mx = np.max([y,yp])
plt.scatter(y,yp)
plt.xlabel('Y_observed_[mg/L]')
plt.ylabel('Y_predicted_[mg/L]')
plt.plot([0, mx], [0, mx], color='k', linestyle='—', linewidth
    =1)
res.summary()
```

☐ Figure 9.7 shows a plot that compares the observed titers with the predicted titers based on the fitted linear regression model in Code 9.8. ☐ Tables 9.6 and 9.7 show some of the information that is output by the OLS test. ☐ Table 9.6 provides a summary of the overall test and some standard statistics that are used in linear regression like the R-squared. ☐ Table 9.7 provides details about the individual effects of each factor. As already discussed, levels for categorical variables appear as individual variables except for the last one of each category. In the columns, we have the regression coefficients, standard errors, t statistic and associated p-value. The p-values can help us here in order to identify which factors had a significant effect. As usual, we will consider the effect of a variable as significant if the p-value< 0.05. The coefficients, in turn, will indicate if there is a positive or negative effect. The following initial conclusions can be drawn from the table:

1. **Origin of replication**: The effect of psC101 is significant (p-value< 0.05) and the coefficient is negative;
2. **Chassis**: The effect of selecting DH5α is not considered significant, which is something that was already observed in ☐ Fig. 9.5. The effect of selecting MG1655, on the other hand, appears as highly significant with a strong positive effect;
3. **Media**: The effect of selecting the growth medium is not conclusive as none of the variables had a significant effect.

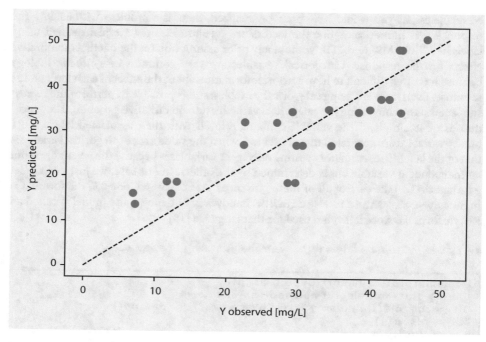

□ Fig. 9.7   Observed vs. predicted titers from the fitted regression model

□ Table 9.6   Output summary of the ordinary linear regression test

Dep. Variable:	y	R-squared:	0.713
Model:	OLS	Adj. R-squared:	0.633
Method:	Least Squares	F-statistic:	8.947
Date:	Wed, 03 Oct 2018	Prob (F-statistic):	0.000208
Time:	10:57:50	Log-Likelihood:	-79.182
No. Observations:	24	AIC:	170.4
Df Residuals:	18	BIC:	177.4
Df Model:	5		

Based on the previous analysis, we arrive at some **redesign rules** that will be employed in the next design/build/test/learn round:

1. **Origin of replication**: select pBR322;
2. **Chassis**: select MG1655;
3. **Media**: repeat for LB, M9 and TB.

In conclusion, we started with $2 \times 3 \times 3 = 18$ possible combinations and we ended up with only 3 possible combinations in our experimental redesign, i.e. a 6-fold reduction in the design space. Obviously, we have kept the numbers small in order to make easier to follow

■ **Table 9.7** Effect contribution analysis from the linear regression test

	coef	std err	t	P>|t|	[0.025	0.975]
Intercept	32.1309	3.612	8.895	0.000	24.542	39.720
ori[T.psC101]	-13.3814	3.143	-4.257	0.000	-19.986	-6.777
Chassis[T.DH5a]	-5.0664	3.724	-1.360	0.191	-12.891	2.758
Chassis[T.MG1655]	16.3840	3.847	4.258	0.000	8.301	24.467
Media[T.M9]	0.0332	3.853	0.009	0.993	-8.062	8.128
Media[T.TB]	2.4211	3.745	0.646	0.526	-5.447	10.289

the example, but the same strategy could be applied for a significantly larger scale often leading in most of the cases to a rapid improvement in performance and a large reduction of the search space.

9.4 Machine Learning-Based Pathway Redesign

We have seen in previous ▶ Sect. 9.3 that a simple statistical test of linear regression can help in order to identify the main factors having an impact in our pathway design. However, the complexity associated with the large combinatorial design space and nonlinear behavior of biological systems can challenge the regression approximation. **Machine learning** provides techniques that are appropriate for dealing with complex systems. Machine learning can be applied in order to infer design rules for synthetic biology circuits similar to the ones that we obtained with linear regression but from rather more complex relationships between the response variable and the design factors. For that very good reason machine learning applied to engineering biology is a very promising approach for tackling current design challenges. At the moment, the field is growing very quickly thanks to the wide availability of powerful, scalable and efficient algorithms and community-driven open source tools. The main challenge is how to efficiently extract information and how to learn from the big data sets that are generated by genomics, transcriptomics, proteomics, metabolomics and other "omics" data.

The aim of machine learning is to give computer systems the ability to learn. In the previous section we used regression analysis in order to learn design rules. Ideally, this task could be delegated to a computer that will automatically apply multiple algorithms in order to find the best predictive model for improving the performance of the metabolic pathway. There exist many machine learning algorithms and strategies available. Discussing all the details of them would involve a full textbook in itself.

In this last section, we will briefly discuss some approaches that have become very popular based on **Bayesian classifiers** as well as those based on **neural networks** and **deep learning**. Neural networks can be applied virtually to any problem in order to build a predictive model. For instance, recent developments on neural networks are making possible the building of models that can learn and predict protein sequence-function relationships at high level of accuracy, surpassing in some cases the innovation capabilities that are found in nature [6, 11].

A word of warning: Mastering the advanced concepts and methods presented here would require some expertise on machine learning algorithms. As we cannot go into full details for every concept used in this section, this introductory part on machine learning will focus on discussing some basic examples in order to provide the reader with a grasp of the capabilities of machine learning in metabolic pathway design. References are given throughout this section as well as at the end of the chapter allowing further reading on the topic.

There are basically two types of problems where we can apply machine learning[15]:

1. **Classification**, where the samples of the training set are labeled, for instance low producer/high producer, and the predictive model should be able to correctly classify new samples;

2. **Regression**, where some dependent variable is measured for each sample, for instance titers of the target compound, and the model should be to predict the value for new samples.

The machine learning procedure starts with a **training set**, which is a data set containing the following information:

- **The input set**: a two-dimensional array or matrix of size $[n \times m]$ corresponding to:
  - n samples: A sample can be either some experimental sample that was generated in the lab or instances extracted from databases;
  - m features: The features are attributes or traits that can be used to describe each item in the sample in a distinctive way;
- **The output set**: a vector or array containing the observed values of the dependent variables. For instance, in ▶ Sect. 9.3 the output set is the observed titers shown in ◻ Table 9.5.

In all cases, the first step is to provide a meaningful representation of the input using the features. Machine learning generally requires quantitative values, either real, discrete or Boolean. However experimental data in synthetic biology often contain very heterogeneous information. All this different information has to be encoded somehow into a quantitative feature matrix. Let's discuss briefly what strategies are available in order to convert the information contained in the experimental data into something that can be fed into the machine learning algorithms:

- **Quantitative variables**: such as temperature, optical density, expression levels, etc. can be directly used as *features*, i.e., columns in our training set;
- **Categorical variables**: such as promoter type, chassis strain, etc. need to be converted into quantitative variables. To that end, we can use the same strategy that was described in ▶ Sect. 9.3, i.e., to define as many features as labels are in the category, each one corresponding to a binary variable that is set to 1 for the samples with the label. This type of representation is known as **one-hot encoding** [1];

---

15 We will discuss here only cases involving supervised learning, i.e., we have data sets that contain both features and labels or output values.

- **Qualitative variables**: like those having discrete levels such as low, medium, high, etc. In this case we can opt either for assigning discrete quantitative values (they can be just their order of activity from lower to higher) or we can use again the one-hot encoding strategy. This is especially useful when we are not sure about the perceived order of preference. For instance, we can assume that growing the culture in rich media corresponds to some high effect, say 10.0 in arbitrary units (a.u.), and that growing in minimal media will correspond to a low value, say 1.0 a.u. However, assigning this type of perceived values is more difficult when the number of possibilities increases and we do not have an a priori clear idea of their influence. Trying to answer questions like should I assign a high number to TB media, a medium number to M9 and a low number to LB are probably biased and meaningless. We could always make some guess based on a quick analysis of the effects as described in ▶ Sect. 9.3. However, a safer approach would be to treat each condition, i.e. TB, M9, and LB, as an independent binary variable using one-hot encoding and to let the machine learning training algorithm to figure out their effects and interactions;
- **Chemical information**: coming from molecular structures need to be converted into features. One possibility is to represent the chemicals through their physico-chemical, molecular, etc. properties such as molecular weight, solubility, number of aromatic rings, druglikeness, etc., as was discussed in ▶ Sects. 4.2 and 6.1. Perhaps more interestingly, we can use an approach based on fingerprints, i.e., long binary vectors where a large number of features describing the molecule are mapped, as discussed in ▶ Sect. 5.2. In both cases, we are talking about feature vectors of fixed length.[16] Another possibility that recently has received a lot of interest from the community [16] is to feed the learning algorithm directly with the SMILES string representation (▶ Sect. 4.2). This type of non-fixed length approach can be used with some types of deep learning algorithms and will be explained later;
- **Protein information**: generally comes in the form of nucleotide or amino acid sequences. Each position in the sequence can be represented using again a one-hot encoding so that each nucleotide position is converted into 4 binary columns (A,C,G,T) or each residue position in the amino acid sequence is converted into 20 binary columns (one per amino acid). Each resulting feature is only set to one when the corresponding nucleotide or amino acid occurs in the sample, respectively. The approach can be used for fixed regions of the sequence, for instance if we are trying to model the effect of a ribosome binding site or a promoter, we can focus in the region around the RBS or promoter sequence. In a more general case, we could take the full protein sequence in order to train a machine learning algorithm. As discussed before in the case of chemicals represented through SMILES strings, this will imply input vectors of variable length and therefore we will need to use some especial type of machine learning algorithms that accept sequences as input rather than fixed-length vectors.

For instance, the training set shown in ▣ Table 9.5 is read again into a pandas dataframe as in Code 9.7. Because the input variables are all categorical, they are transformed using the one-hot encoding. As shown in Code 9.9, pandas provides an easy way to perform the encoding using the get_dummies method (variables resulting from a one-hot encoding are often called "dummy" variables) (▣ Table 9.8).

---

16 Fingerprints are defined as vectors of fixed length through hashing (see ▶ Sect. 5.2).

◘ **Table 9.8**   One-hot encoding of the training set

DH10b	DH5a	MG1655	LB	M9	TB	pBR322	psC101	y [mg/L]
1	0	0	1	0	0	0	1	28.44
0	1	0	0	1	0	1	0	22.26
0	1	0	1	0	0	0	1	7.15
1	0	0	1	0	0	0	1	12.96
0	0	1	1	0	0	1	0	44.22
0	1	0	1	0	0	1	0	30.45
0	0	1	0	1	0	1	0	43.81
1	0	0	0	0	1	1	0	44.10
0	1	0	1	0	0	1	0	34.44
1	0	0	0	1	0	0	1	11.57
1	0	0	0	0	1	1	0	28.99
1	0	0	0	1	0	0	1	29.49
0	1	0	0	0	1	0	1	6.85
0	0	1	1	0	0	0	1	34.06
0	1	0	0	1	0	1	0	38.32
0	0	1	0	0	1	0	1	41.34
0	1	0	1	0	0	1	0	29.60
0	0	1	0	0	1	1	0	47.69
0	0	1	0	1	0	0	1	39.81
0	0	1	0	0	1	0	1	42.48
1	0	0	0	1	0	1	0	22.37
1	0	0	0	0	1	1	0	38.25
0	1	0	0	0	1	0	1	12.21
1	0	0	1	0	0	1	0	26.84

■■ Code 9.9 A `pandas` **dataframe** `df` **with the experimental values and labels in**
◘ Table 9.5 **is transformed into one-hot encoding using the** `get_dummies`
method.

```
df = pd.read_csv('table.csv')
dfd = df.get_dummies( df )
display( dfd )
```

The next step is to analyze and visualize the data using techniques such as principal component analysis (PCA). Such analysis focuses on finding linear combinations of features where most of the variability can be found. They can be used in order to reduce the number of features. In our example, which uses dummy variables with a low number of samples coming from an experimental result such type of analysis is not necessary, but in more general cases, especially the ones using quantitative independent variables, PCA analysis as well as other feature selection techniques can help reducing the dimension of the input space.

Multiple machine-learning algorithms for classification and regression exist. We are going to train a supervised classifier for our example. The two labels are "low producer" and "high producer". Based on ◘ Fig. 9.7, the threshold in the titers is defined as $y \geq 20$ mg/L as it can be seen a good separation between the two classes for that threshold. Perhaps the simplest classifier is a **naive Bayes algorithm** which is based on applying the Bayes theorem with the assumption of independence between features. The procedure is shown in Codes 9.10 and 9.11. The score of the classifier of 0.92 is the accuracy, i.e., the number of correctly predicted samples divided by the total number of samples. Therefore, in this example a good fitting of the data to a naive Bayes classifier was obtained.

■■ Code 9.10 Fitting of a naive Bayes classifier from the training set.

```
from sklearn.naive_bayes import GaussianNB
gnb = GaussianNB()
Yc = Y > 20
gnb.fit(X, Yc)
print( gnb.score(X, Yc). round(2) )
```

■■ Code 9.11 Fitting accuracy score for the naive Bayes classifer in Code 9.10.

```
0.92
```

Evaluating the performance of fitting a model is an initial way of assessing the quality of our predictions. However, there is the risk of overfitting, which means that the model is able to reproduce the training set but the performance deteriorates when trying to predict new data. In order to avoid such issue, we generally perform a cross-validation of the model: the data is split into train and test set. For instance, Codes 9.12 and 9.13 show how to perform such test using the `train_test_split` function. In this example, 25% of the data, i.e. 6 points, are randomly selected for the test set. In the test, 5 samples in the test set were high producers and 1 was low producer. The low producer was correctly classified, however one of the high producers was misclassified as low producer. That means a precision of 1.0 for high producers (4 of the predicted 4 were correctly retrieved) and of 0.5 for low producers (1 of the predicted 2 was correctly retrieved), i.e., a 0.92 average precision, and a recall of 1.0 for low producers (1 of the total 1 low producers was retrieved) and 0.80 for high producers (4 of total 5 high producers were retrieved), which makes an average recall of 0.83.

■■ Code 9.12 Splitting the data into train and test sets.

```
from sklearn.model_selection import train_test_split
from sklearn import metrics
X_train, X_test, y_train, y_test = train_test_split(X, Yc,
    test_size=0.25)
gnb = GaussianNB()
gnb.fit(X_train, y_train)
y_predict = gnb.predict(X_test)
print( metrics.classification_report(y_test, y_predict))
```

■■ Code 9.13 Output scores from cross-validation in Code 9.12.

	precision	recall	f1−score	support
False	0.50	1.00	0.67	1
True	1.00	0.80	0.89	5
avg / total	0.92	0.83	0.85	6

Performing the previous test only once would not provide enough evidence about the actual performance of the classifier. This is why the most common approach is to repeat multiple times the test by performing an *n-fold cross-validation*, where the data is randomly split into *n* subsets, each one is used as a test set. In that way, we obtain several measurements of the performance and we can evaluate the overall performance by averaging. Codes 9.14 and 9.15 show an example of an *n*-fold cross-validation using cross_val_score with a score (accuracy) of 0.92 ± 0.1.

■■ Code 9.14 *n*-fold cross validation of the naive Bayes classifier.

```
from sklearn.model_selection import cross_val_score
gnb = GaussianNB()
scores = cross_val_score(gnb, X, Yc, cv=5)
print( np.mean(scores).round(2),np.std(scores).round(1) )
```

■■ Code 9.15 Output mean and standard deviation accuracy score for the *n*-fold cross validation of the naive Bayes classifier in Code 9.14.

```
0.92  0.1
```

The previous example was a simple illustration of the use of a machine learning algorithm (the naive Bayes classifier) in order to build and validate a predictive model. In a real scenario, we are mainly interested in applying machine learning to highly complex problems involving many input features and big amounts of experimental data. As a matter of fact, the most successful examples of the use of machine learning in biotechnology always involved large training sets, for instance the full set of reactions in chemical databases [16], the full set of protein sequences [10], gene expression data [4], etc.

One of the main approaches for dealing with **big data** is **deep learning** [5]. Deep learning is a subfield of machine learning where the learning is performed through successive *layers*, each one intended to encapsulate some meaningful representation of the data. The way models are built in deep learning is through **neural networks**, which are a type of machine learning algorithms that try to reproduce in a more or less loosely way the learning mechanisms of the brain. A neural network is composed of multiple interconnected nodes (they can be thousands) organized in layers and some algorithm such as backpropagation is used in order to adjust the weights of the connections between the nodes. Moreover, each node has an activation or transfer function so that the input information is transformed before sending it to the next layer. The advantage of neural networks is that they can approximate highly complex non-linear relationships between the input features and the dependent variables. They can also be retrained with new data as them become available in order to refine the model.

There exist many types of neural networks, often called architectures, each one serving a different purpose. Modern neural networks are designed to operate with tensors, which are a mathematical representation for highly multidimensional data. Popular software for deep learning are `TensorFlow`,[17] `Theano`,[18] and `PyTorch`.[19] A good way to start working with deep learning is by using the high-level and user-friendly Python library `Keras`.[20] The paradigm of deep learning is that we can stack different layers of neural networks with different architectures in order to obtain an efficient internal representation of the information, what sometimes is called the **latent space**. In the context of pathway design, an interesting advantage of using neural networks and deep learning is that there are some architectures that have been designed with the goal of dealing with *sequential input data*. Such sequential data could be the time-course response of some variable, the pixels in an image, the sequence of words in some text, etc. Instead of having to convert that signal into some basic features like average, deviation, etc., we can feed the sequence directly into the neural network. This can be achieved with **recurrent neural networks** (RNNs) such as the long-short term memory (LSTM) algorithm. Discussing such algorithms in detail is out of the scope of this textbook, but the reader can find some introductory references at the end of this chapter.

Protein sequences or string representations of chemicals can be used as inputs to RNNs, avoiding in that way the always challenging decision of focusing on some specific region of the sequence or selecting a reduced set of chemical descriptors. **Autoencoders** are a class of architectures of neural networks that consist of an input RNN (the encoder) and an output RNN (the decoder). Such networks are designed to learn to accurately reproduce the input sequence from an internal non-sequential (generative) information-rich representation, the latent space. The resulting latent space representation can potentially be used to predict changes in the sequences of the genetic circuits having a positive impact in an engineered metabolic pathway. Machine learning is therefore expected to play a prominent role in metabolic pathway design by learning complex relationships between the biological parts, delivering in that way efficient engineering strategies for improving production.

---

17 ▶ https://www.tensorflow.org/
18 ▶ http://deeplearning.net/software/theano/
19 ▶ https://pytorch.org/
20 ▶ https://keras.io/

┌─ **Take Home Message** ───

— Metabolic pathway optimization is an iterative process that requires of several
  rounds of the design/build/test/learn cycle.
— Pathway redesign can be performed through two different approaches: model-
  driven and data-driven.
— Model-based pathway redesign can be achieved by
  1. Performing genome-scale simulations in order to select optimal growth
     conditions and best gene deletions;
  2. Computational enzyme design in order to improve performance of the enzyme
     bottlenecks.
— Data-driven pathway redesign can be achieved by
  1. Linear regression analysis of the experimental data from a combinatorial design
     in order to identify effects having the highest impact;
  2. Machine learning algorithms trained on the experimental data in order to build
     predictive models that consider complex interactions between the design
     factors.

9.5 Problems

9

? **9.1 Media optimization.**

✓ Repeat the media optimization procedure shown in Code 9.2 by varying the following
growth conditions: (a) Phosphate exchange (`EX_pi_e`); (b) Ammonia exchange (`EX_nh4_e`); (c) CO_2 exchange (`EX_co2_e`); (d) H_2O exchange (`EX_h2o_e`); (e) H_2
exchange (`EX_h2_e`). Can you see any improvement in the strain performance? *Hint:*
Some of the exchange fluxes need to be of net consumption (negative) in order to
make the cell growth viable, while others are of net production (positive). By looking at
the optimal range of fluxes through flux variability analysis (see ► Sect. 2.4) for the
default conditions, you can determine the preferred direction of the exchange fluxes.

? **9.2 Chassis optimization.**

✓ Set the objective function of the cell to several combinations of growth and target
production: $\texttt{Objective} = \alpha \texttt{Biomass} + (1-\alpha)\texttt{Production}; (0 \leq \alpha \leq 1)$ and perform
single gene deletion. Do the target mutations improving the objective change in
function of α? Optionally, you can try double gene mutations for some of the top
predicted knock-outs, as in Code 9.6.

? **9.3 Enzyme redesign based on database information.**

✓ ▫ Tables 6.4 and 6.5 contain the information about the 4 enzymes involved in the
pinocembrin pathway. For each enzyme, look for the following information: (a) Go to
the Catalaytic Site Atlas (CSA) database[21] and retrieve the list of residues that

───

21 ► http://www.ebi.ac.uk/thornton-srv/databases/CSA/

reported with a catalytic function; (b) Go to `Brenda` database and look in Molecular Properties for information about protein engineering, compile a table with all known single or multiple mutations having a positive effect on the activity. Is there any overlap between the residues in CSA and the mutations in Brenda? Can we infer some rule about the relationship between the catalytic region and the location of the mutations? *Hint:* Look at the location of the catalytic site and the mutations in the 3D structure by using some molecular visualization software like `pyMOL`[22] or `Chimera`.[23]

? 9.4 **Enzyme redesign based on predictions.**

✓ Based on the information about catalytic sites and mutants obtained from databases in the previous Question 9.3 for the enzymes in the pinocembrin pathway, (a) select a *hot region*, i.e., a region containing important functionally-related residues, comprising 10–15 residues in the 3D structure; (b) submit the structure to `FuncLib`[24] and select for the resulting predicted mutants a set of candidate pairs of mutants. Do you see any significant increase in performance in the predictions?

? 9.5 **Assessing the effect of interactions.**

✓ In ▶ Sect. 9.3, the effects of each factor assuming a model with the formula: `y~ori+Chassis+Media` was assessed. Such model does not take into account higher order interactions, i.e., the linear model assumes that the effect of each factor is independent from the others. However, we may often find that the join effect of two or more factors has a significant effect. This can be tested by using a procedure similar to the one in Code 9.8 but using a formula that takes into account second order interactions:

`y~ori+Chassis+Media+ori*Chassis+ori*Media+Chassis*Media.`

Repeat the ordinary linear regression test with the given formula and determine if any of the interactions between the factors appears as significant.

? 9.6 **Training a predictive model using machine learning.**

✓ ◻ Table 8.7 is a combinatorial design of 24 constructs involving 11 independent factors. In ◻ Table 9.5, only 3 factors were considered, assuming the effect of the rest to be negligible. Using the same procedure as in ▶ Sect. 9.4, build naive Bayes classifiers that consider the following additional effects from the original design: (a) selection of multiple PAL enzyme sequences; (b) selection of PAL promoter; (c) selection of 4CL enzyme sequence; (d) selection of 4CL promoter. How the performance of the models change by adding additional input factors?

References

1. Angermueller, C., Prnamaa, T., Parts, L., Stegle, O.: Deep learning for computational biology. Mol. Syst. Biol. **12**(7) (2016)
2. Bartsch, S., Bornscheuer, U.T.: Mutational analysis of phenylalanine ammonia lyase to improve reactions rates for various substrates. Protein Eng. Des. Sel. **23**(12), 929–933 (2010). https://doi.org/10.1093/protein/gzq089
3. Buß, O., Rudat, J., Ochsenreither, K.: FoldX as protein engineering tool: better than random based approaches? Comput. Struct. Biotechnol. J. **16**, 25–33 (2018). https://doi.org/10.1016/j.csbj.2018.01.002
4. Chen, Y., Li, Y., Narayan, R., Subramanian, A., Xie, X.: Gene expression inference with deep learning. Bioinformatics **32**(12), 1832–1839 (2016). https://doi.org/10.1093/bioinformatics/btw074
5. Chollet, F.: Deep learning with Python. Manning Publications (2018). https://www.manning.com/books/deep-learning-with-python
6. Colwell, L.J.: Statistical and machine learning approaches to predicting protein-ligand interactions. Curr. Opin. Struct. Biol. **49**, 123–128 (2018). https://doi.org/10.1016/J.SBI.2018.01.006
7. Holm, L., Laakso, L.M.: Dali server update. Nucl. Acids Res. **44**(W1), W351–W355 (2016). https://doi.org/10.1093/nar/gkw357
8. Khersonsky, O., Lipsh, R., Avizemer, Z., Ashani, Y., Goldsmith, M., Leader, H., Dym, O., Rogotner, S., Trudeau, D.L., Prilusky, J., Amengual-Rigo, P., Guallar, V., Tawfik, D.S., Fleishman, S.J.: Automated design of efficient and functionally diverse enzyme repertoires. Mol. Cell (2018). https://doi.org/10.1016/J.MOLCEL.2018.08.033
9. Li, W., Godzik, A.: Cd-hit: a fast program for clustering and comparing large sets of protein or nucleotide sequences. Bioinformatics **22**(13), 1658–1659 (2006)
10. Li, Y., Wang, S., Umarov, R., Xie, B., Fan, M., Li, L., Gao, X.: DEEPre: sequence-based enzyme EC number prediction by deep learning. Bioinformatics (2017). https://doi.org/10.1093/bioinformatics/btx680
11. Mura, C., Draizen, E.J., Bourne, P.E.: Structural biology meets data science: does anything change? Curr. Opin. Struct. Biol. **52**, 95–102 (2018). https://doi.org/10.1016/J.SBI.2018.09.003
12. Ortiz, A.R., Strauss, C.E., Olmea, O.: MAMMOTH (matching molecular models obtained from theory): an automated method for model comparison. Protein Sci. **11**(11), 2606–2621 (2002). https://doi.org/10.1110/ps.0215902
13. Richter, F., Leaver-Fay, A., Khare, S.D., Bjelic, S., Baker, D.: De novo enzyme design using Rosetta3. PLoS One **6**(5), 1–12 (2011). https://doi.org/10.1371/journal.pone.0019230
14. Samish, I.: The framework of computational protein design. Methods Mol. Biol. (Clifton, N.J.) **1529**, 3–19 (2017). Springer. https://doi.org/10.1007/978-1-4939-6637-0_1
15. Schumacker, R.E.: Learning statistics using R. Sage Publications, Thousand Oaks (2014)
16. Segler, M.H.S., Preuss, M., Waller, M.P.: Planning chemical syntheses with deep neural networks and symbolic AI. Nature **555**(7698), 604–610 (2018). https://doi.org/10.1038/nature25978
17. Wijma, H.J., Fürst, M.J.L.J., Janssen, D.B.: A computational library design protocol for rapid improvement of protein stability: FRESCO. Methods Mol. Biol. **1685**, 69–85 (2018). Humana Press, New York. https://doi.org/10.1007/978-1-4939-7366-8_5

Further Reading

A discussion about how **computational protein design** is expanding synthetic biology applications:
Gainza-Cirauqui, P., Correia, B.E.: Computational protein design – the next generation tool to expand synthetic biology applications. Curr. Opin. Biotechnol. **52**, 145–152 (2018). https://doi.org/10.1016/J.COPBIO.2018.04.001

The site https://www.scipy-lectures.org/ provides a good introduction to many topics related to scientific computation using Python. The sections about statistics and machine learning are a useful reference material for the methods discussed in this chapter.

An excellent introduction to **deep learning** using Python:
Chollet, F.: Deep learning with Python. Manning Publications (2018).

Supplementary Information

© Springer Nature Switzerland AG 2019
P. Carbonell, *Metabolic Pathway Design*, Learning Materials in Biosciences,
https://doi.org/10.1007/978-3-030-29865-4

Appendix A

The Biodesign Toolbox

Metabolic pathway design is an engineering technology mainly assisted by computational tools. Pathway design leverages computer-aided design in order to create new genetic constructs. The modern bench desk of a metabolic pathway designer makes use of the best technologies available in order to engineer biology. At the time of this writing, Python is the best technology for metabolic pathway design. Therefore, this book requires some knowledge about Python. Some examples use simple Python coding, others use more complex methods that take advantage of the increasingly large selection of Python packages for synthetic biology. Hopefully, you will find most of the examples insightful. I encourage you to write your own code, the examples are just starting suggestions but the world of synbio Python is vast and the possibilities are infinite.

The first thing that you need to do is to install your own Python. I would recommend for this using Anaconda[1] because is cross-platform, i.e., what you did in your Windows computer would also work in Linux and OS X. Linux is by far the best test bench computational platform for synthetic biology. It is science-oriented and can be tailored to your needs. Working on a Linux virtual machine is a great option, many cloud services like Google or Amazon provide you the possibility of starting your virtual Linux machine. The future of synbio design will probably be based on Linux, so I strongly recommend that you learn its basics as soon as possible.

In this textbook, we have carefully chose those Python libraries that are essential for your work. As a matter of fact, there are many other libraries that are useful for synbio projects. Every day, hundreds of new libraries are created and shared. Therefore, do not shy away from install-ing and testing others than the ones in this textbook. Check regularly the literature and GitHub[2] for new packages.

Once you have your Anaconda system install, the next thing is to create a computing environment. An environment means a computational system that has all required libraries and packages in order to run your project. In our case, our project is metabolic pathway design. We will use Python 3.5, because it seems the most stable version for pathway design at the time of this writing, as shown in Code A.1.

Code A.1   Create a new coding environment.

```
conda create —name pathdes python=3.5
conda activate pathdes
# source activate pathdes : in earlier
  versions of conda
```

After this operation you are in a new environment where to install the scientific software. For instance Code A.2 will install the RDKit chemoinformatics library using the rdkit channel.

Code A.2   Create a new coding environment.

```
conda install
—c rdkit rdkit
```

The basic tools that you need to install are listed in ◻ Table A.1. These libraries are community-based basic tools that will help you to develop research at the highest-quality levels. All of these packages have well documented web pages. I recommend the reader to go through the sites in order to learn the features of these great packages. Eclipse[3] is one of the best integrated environments for programming, Spyder[4] is an

---

1   ▶ https://www.anaconda.com/

2   ▶ https://github.com/
3   ▶ https://www.eclipse.org/
4   ▶ https://www.spyder.com/

▣ **Table A.1**   Summary of main Python synbio packages

Package	Description
numpy	Add support for multidimensional matrices
scipy	Scientific and technical computing
pandas	Data manipulation and analysis
jupyter	Interactive data science and scientific computing
matplotlib	Plotting library
seaborn	Data visualization
cobrapy	Constrained-based genome-scale models simulation
biopython	Bioinformatics in Python
pysbol	Synthetic biology SBOL tools
scikit-learn	Statistical machine learning
keras	Deep learning

environment focused on scientific programming. Jupyter notebooks[5] are a convenient environment for testing and presenting your code using a web-based interface. Some free web-based services like Collaboratory[6] offer the possibility of creating, running and sharing Jupyter notebooks in the cloud.

I have setup a Github site[7] in order to share our ideas and projects on metabolic pathway design. You are welcome to join the community.

5 ▶ http://jupyter.org/
6 ▶ https://colab.research.google.com

7 ▶ https://pablocarb.github.io/PathwayDesign/

Index

Printed in the United States
By Bookmasters